Praise for *Supply Shock*

It may be premature to call this book a masterpiece, but it's evident that Czech has mastered the art of melding science, economics, policy and politics in one readable piece. *Supply Shock* belongs in the classroom, boardroom, town halls and policy circles. It belongs in the hands of all those who care, as Czech might say, "about the grandkids."
—Herman Daly, Professor Emeritus, University of Maryland, School of Public Policy; author of *Steady State Economics*; Lifetime Achievement Award winner, National Council for Science and the Environment

An old economic world is dying, and a new economic world is being born. Brian Czech is one of the midwives of this new economic world.
—Governor Richard D. Lamm

This is a brave book that raises questions we all need to ask and try to answer. Czech proposes the evolution of a revolution, thinking and feeling and working our way toward a fair, sustainable, constructive social order in America and all around the world. The style is clear, cogent, honest, stimulating, free of clutter, and often amusing; it's boredom-free. You'll enjoy it.
—Neil Patterson, president, Neil Patterson Productions; past president, W.H. Freeman and Company, co-founder of Benjamin-Cummings, Worth, and Scientific American Books

Supply Shock clearly describes the heart of what ails us—a zombie-like addiction to economic growth everywhere at all costs. Brian Czech brilliantly dissects the economic theories, models, and mindsets that are diminishing the human prospect while calling it 'progress'. . . . King Midas would have understood the point, as we will someday. There are biophysical limits to economic and population growth and we ignore them at our peril.
—David W. Orr is Paul Sears Distinguished Professor of Environmental Studies and Politics and Senior Adviser to the President, Oberlin College; author of seven books; Lyndhurst Prize winner

Brian Czech has used a remarkable combination of education and experience to build a solid reputation as an innovative thinker. As a wildlife biologist, wilderness ranger, and natural resources advisor to Native American tribes, Czech developed a keen awareness of the status and trends of the American landscape. Then, with graduate studies in political science and post-grad studies in economics, followed by years as a conservation biologist and planner in a federal natural resources agency, Czech put the pieces together to envision an ecologically and economically sustainable future. His are not the loosely-framed and impractical solutions of a casual dreamer or a politically naive zealot. *Supply Shock* is the offering of a man who has tested his ideas, exposed them to peers and colleagues, and appears at countless meetings and conventions where he defends his convictions. *Supply Shock* is an adventure in learning. Czech's vision of "steady statesmanship" is impressive and convincing, and this book easily qualifies as one of the key manuals for those who care about the world and its inhabitants.
—Lynn Greenwalt, former director, U.S. Fish and Wildlife Service

Dr. Brian Czech has dedicated his entire professional life towards the study of wildlife conservation, environmental protection, and human society. *Supply Shock* is the culmination of this thinking, and should be read by leaders, and upcoming professionals in natural resource conservation and environmental management. Bold leadership – the kind needed for management and conservation of the world's natural resources and habitats—can be enhanced by Czech's vision of steady statesmanship.
—Paul R. Krausman, Boone and Crockett Professor of Wildlife Conservation, University of Montana, and past president, The Wildlife Society

The practice of conservation biology has a palpably futile feeling when economic growth is the summum bonum. *Supply Shock* provides an antidote. All who are serious about the big picture of biodiversity conservation should read this book. It will change your idea of what the future can be, and how to create that future.
—Paul Beier, president, Society for Conservation Biology, and Regents' Professor, School of Forestry, Northern Arizona University

Brian Czech comes to the rescue with an honest look at what the global economy is really doing to the earth as he challenges the cherished goal of economic growth. Many who write on big economic ideas lack a deep knowledge of the amazing interactions of the forms of life on our planet and their relevance to economic analysis. *Supply Shock*, in contrast, brings together the keen observations of a skilled biologist with a deep understanding of our failing economic system. Brian Czech has come up with the major economic rethinking needed to prevent cascading collapses of human societies and the rest of the species on the planet.

—Brent Blackwelder, Past President, Friends of the Earth; Founding Chairman, American Rivers

The past century of explosive population and economic growth, a period that people today take to be the norm, is actually the single most anomalous period in human history—and it threatens to do us in! Growth is normally just the juvenile phase of the life cycle. With maturity, growth slows but development continues as living things become better adapted to their socio-ecological contexts. In *Supply Shock*, Brian Czech graphically shows how the growth-based status quo is destroying the ecological basis of human existence and eloquently describes an alternative path to true economic maturity. A dynamically-evolving but non-growing steady-state economy offers humanity's best hope for achieving a just and secure sustainability within the means of nature.

—Bill Rees, author, Professor Emeritus of Human Ecology and Ecological Economics, University of British Columbia School of Community and Regional Planning, and co-winner of the 2012 Boulding Prize in Ecological Economics and a 2012 Blue Planet Prize.

This well-written and comprehensive volume is a great resource for the issue of questioning "economic growth" and beginning to think about how to move towards a new paradigm for the earth's future. For a society that is trapped in mode of continued growth as a necessity, much like a person riding on the back of a hungry tiger, we need all the help we can get to find our way to a sustainable economic model.

—Doug La Follette, Secretary of State, Wisconsin

Brian Czech marries economics, biology, and political science in a brilliant account of why we need to abandon growth and build a new governance system. There is no sociable alternative to the steady state economy.

> —Lorenzo Fioramonti, Jean Monnet Chair in Regional Integration and Governance Studies at the University of Pretoria; Senior Fellow at the Centre for Social Investment, University of Heidelberg; author of numerous books on international politics and governments, including *Gross Domestic Problem: The Politics Behind the World's Most Powerful Number*

Economic growth is so 20th century. Remember cheap oil, rural electrification, and Mad Men? They gave us history's biggest hit of expansionary exuberance. But today what little growth we see comes from consumer debt, deficit spending, and natural resource liquidation. This can't go on, and it won't. What's the alternative? As Brian Czech lucidly explains, it's time for our economy to start acting like a responsible adult in a world of limits. This book reeks of sanity: read it!

> —Richard Heinberg author, *The End of Growth*

SUPPLY SHOCK

ECONOMIC GROWTH AT THE CROSSROADS AND THE STEADY STATE SOLUTION

BRIAN CZECH

Steady State Press
steadystate.org

Copyright © 2013 Brian Czech.

All rights reserved. No part of this publication may be reproduced, distributed, or transmitted in any form or by any means, including photocopying, recording, or other electronic or mechanical methods, without the prior written permission of the copyright holder, except in the case of brief quotations embodied in critical reviews and certain other noncommercial uses permitted by copyright law. For permissions requests, please email the publisher at info@steadystate.org, with the subject, "Permissions Request."

Includes bibliographic references and index.

ISBN: 978-1-7329933-2-7 (Paperback)

Library of Congress Control Number: 2020914163

Cover design by Diane McIntosh. © iStock (jhorrocks)

First printing April 2013 in Canada; New Society Publishers

Second printing August 2021 in USA; Steady State Press

1. Environmental economics. 2. Economic development.
3. Economic history. 4. Economic policy.

Steady State Press is an imprint of the Center for the Advancement of the Steady State Economy (CASSE), www.steadystate.org.

HC79.E5C94 2013   338.9'27   C2013-900654-0

Steady State Press
steadystate.org

CONTENTS

Foreword by Herman Daly ix
Preface xiii

PART 1. ECONOMIC GROWTH AT THE CROSSROADS
1. It Really *Is* the Economy, "Stupid!" 3
2. Good Growing Gone Bad 23

PART 2. THE DISMAL SCIENCE COMES UNHITCHED
3. Classical Economics: Dealing with the Dismal 51
4. "Neoclassical" Economics: Dealing with the Devil . . . 75
5. Not of This Earth 117

PART 3. ECONOMICS FOR A FULL WORLD
6. Ecological Economics Comes of Age 137
7. Don't Sell the Farm: The Trophic Theory of Money . . . 171
8. Technological Progress and Less-Brown Growth 195

PART 4. POLITICS AND POLICY: THE HORSE BEFORE THE CART
9. "What Have You Done for Growth Today?" 225
10. Hummer Haters: The Steady State
    Revolution Revisited 259
11. A Call for Steady Statesmen:
    Policies for a Full-World Economy 275

Notes 329
Literature Cited 347
Index 359
About the Author 367

FOREWORD

by Herman Daly

A steady state economy is the goal that both Brian Czech and I ended up advocating. But the paths by which we arrived at our common destination were different. I saw things as an economist looking from within the economy outward toward its containing ecosystem. I saw the constraints put on economic growth by the fact that the biosphere is finite, non-growing, materially closed and receives a fixed rate of inflow of solar energy. My problem then was to study ecology and try to integrate it with economics.

Czech, as a wildlife ecologist and conservation biologist, looked from the ecosystem inward toward the growing economy and wondered how he and his colleagues could ever conserve ecosystems and species if the economy kept on growing and absorbing into itself ever more of nature. He concluded that his professional goal was doomed to failure in a world dominated by economic growth. His problem then was to study economics and to integrate it with ecology.

Supply Shock is the culmination of Czech's journey, and he's paved the way for generations of ecologists to follow. It is encouraging to me that, given our different starting points, we end up at the same destination.

Czech's self-directed study of economics started with the history of economic thought, learning from the great economists and digesting their fundamental ideas. By this procedure he gleaned a lot of insight—and shares it with the reader—that has escaped recent PhDs in economics whose curriculum usually dropped history of economic thought to make room for more mathematics and econometrics. Of course mathematics needs no defense, but its considerable power comes from abstraction, and the modern economists' excessive pursuit of mathematical formalism led them to abstract from just about everything important—including natural resources! So Czech's concrete focus on material and energy, thermodynamics and trophic levels is a welcome corrective. Just how welcome is evident from his interpretation of Mason Gaffney's thesis on the corruption of economics. It seems all the neoclassical abstracting from land and natural resources had less innocent motives than just mathematical simplification. But I don't want to give away the story!

Czech's roots as a blue-collar country boy turned ranger, biologist and eventually economist come through in his colorful writing style and agrarian metaphors. But it would take a very dull city slicker not to perceive that beneath this rustic exterior is a keen mind honed by years of study in science and economics, as well as by much policy experience and political acumen gained as a longtime civil servant and activist. This was not a book written while on mountaintop sabbatical with foundation backing in pursuit of tenure and promotion. It was financed by the "Czech Foundation," written on weekends, at night and on vacation time, motivated by the fact that the author has something important to say. Thank goodness Czech was determined to pull it off. The book will not win him a promotion in the federal government where growth politics still prevail. But I think it will earn the admiration and recommendation of the many readers who are still able to think for themselves amidst the political and media greenwashing about "win-win" policies for promoting both economic growth and environmental conservation.

Al Gore had the courage to point to "an inconvenient truth." But *the* inconvenient truth is that there are limits to economic growth. Coloring it green or calling it "smart" (as if others favor dumb growth) is at best a palliative. There is even a limit to growth in the number of trees we can plant, species we can preserve and Priuses we can buy. We live in a full world—and full-world economics requires that empty-world economic growth policies be radically changed. Czech deftly handles the issue of limits and offers a wealth of ideas about how to live—and live well—within those limits. His vision of "steady statesmanship" in international diplomacy is alone worth the price of *Supply Shock*. Czech's recent forays into United Nations dialogue give credence to the hope that steady state economics will catch on in international affairs.

Many authors have written about economic growth, the history of growth theory, national income accounting, the nuances of technological progress, ecological economics, the politics of economic growth and the policy solutions of steady state economics. Few have undertaken the daunting task of integrating it all in one book. It's all integrated in *Supply Shock*. It may be premature to call this book a masterpiece, but it's evident that Czech has mastered the art of melding science, economics, policy and politics in one readable piece. *Supply Shock* belongs in the classroom, boardroom, town halls and policy circles. It belongs in the hands of all those who care, as Czech might say, "about the grandkids."

HERMAN DALY is Emeritus Professor at the University of Maryland's School of Public Affairs. He was Senior Economist with the World Bank and has authored over a hundred journal articles and numerous books, including *For the Common Good* and *Beyond Growth*. Daly has received the Honorary Right Livelihood Award (Sweden), the Heineken Prize for Environmental Science (Netherlands), and the Lifetime Achievement Award from the National Council for Science and the Environment (United States).

PREFACE

This book is about that great engine of history that gets presidents elected, assembles armies and sends men to the moon. Here it builds an Eiffel Tower, there it dams a Yangtze River. We find it sending foreign aid one day, only to wage war the next. For better or for worse, it moves mountains, literally and figuratively.

What is this "it" that sounds so omnipotent yet unpredictable? It is, as the historian J. M. McNeil put it, "easily the most important idea of the 20th century." He might have added that it was already a pretty important idea in the 19th century, and certainly is no less important in the 21st. This big idea, this engine of history, this most godlike of government goals, is economic growth. Economic growth holds the most prominent spot in domestic policy matters and arguably in international affairs.

Sadly, for many people the syllables "econ" conjure up such boring memories that serious public dialog about economic growth is like a baby thrown out with the bathwater. This is probably due to the tedious way economics is taught in high schools, colleges and universities. It's a shame, because so much of our world—both good and bad—is linked at the hip with economic growth, and more dramatically by the day. None of us are immune to its effects. They say "a rising tide lifts all boats," but with economic growth we're all in the same boat, navigating a rising tide. In another sense, we do occupy different boats: some are luxury liners, while others are skiffs being thrashed about in their wake. Either way, the seas are rising and we're all at sea.

Economic growth was a good goal during most of human history, meaning it was good for humans in general, no doubt. But the central thesis here is that economic growth has become a bad goal at this point in history, especially in the United States, Western Europe, Japan and other highly developed nations. We are at a crossroads that is not only immensely important socially, it is perhaps the most important crossroads in the history of public policy issues. Politicians and economists who continue to advocate economic growth often mean well but do not understand the implications. They tend to have no background in the sciences most relevant to economic growth at this point in history. Meanwhile, there is an insidious system of government, especially in the United States with its approach to campaign financing, that will tend to uphold the goal of economic growth regardless of its merits.

Yet most citizens are starting to get the sense that something is amiss. Common sense and general experience tell them that something just doesn't square with the political rhetoric that "there is no conflict between growing the economy and protecting the environment." At the same time, more citizens are seeing that their own grandkids' *economic* welfare depends on us protecting the environment today. Few things demonstrated this as ruthlessly as British Petroleum's Deepwater Horizon oil spill, which threatened many and stole some potential jobs of future shrimpers, oystermen, and a whole chain of service sector workers dependent upon healthy fisheries in the Gulf of Mexico.

Could it be that something is wrong with the sheer immensity of our national and global economies? Of course it could, and the sooner we recognize it the better. Fortunately there is a clear, realistic and sustainable alternative to economic growth that citizens and consumers can demand and attain. It's an economy that neither grows nor shrinks, within reasonable bounds. It's called a "steady state economy," and this prospect should give us hope and courage in a world gone crazy on growth. We can demand an end to economic growth and pursue the establishment of a steady state economy. We should demand it first in the United States, Europe and some Asian countries where we can most afford it, then in the rest of the world.

What does this mean, "demand?" First, it does not mean a communist revolution or an armed insurrection of any type, nor even vandalism, much less any acts of terrorism. Rather, we can demand the steady state economy peaceably in our social relations, our political activities and with our preferences in the market.

We cannot claim to know the precise sizes these local, regional and national economies should take, but the time is now to stop our wealthier economies from further bloating. This cannot happen overnight, but it is time to apply the brakes, and firmly. We must risk some skidding and maybe some injuries to avoid a fatal crash. In fact, the global economy will probably have to *shrink* before a steady state can fit on the planet, and many European scholars are uniting with activists under the banner of "La Décroissance."[1] For purposes of equity and political stability, they say this global process must include a period of economic degrowth in the wealthiest economies and a period of economic growth in the poorest, but with a net effect of shrinkage. Almost surely they are right, too, but the major paradigm shift necessary at this point in history is away from economic growth, and ultimately the steady state economy remains the only sustainable long-term policy goal.

We're at a crossroads, alright. We're in a world of climate change, financial crises, economic meltdowns, biodiversity collapse, resource shortages and environmental calamities. What we are facing is no temporary, localized "supply shock" to be absorbed by the larger economy. This isn't a seasonal water shortage or a spike in the price of bacon. This is the mother, macroeconomic Supply Shock to the global economy and all its constituent nations. The biggest idea of the 20th century has led to the biggest problem of the 21st. Are we ready?

PART 1

ECONOMIC GROWTH AT THE CROSSROADS

CHAPTER 1

It Really *Is* the Economy, "Stupid!"

> *We should* double *the rate of growth,*
> *and we should* double *the size*
> *of the American economy!*
> JACK KEMP

QUICKLY AND OMINOUSLY, bottles of drinking water have appeared on grocery shelves all over the world.[1] Remember, it wasn't that long ago when a bottle of water was a novelty for a grocery store. It wasn't too surprising to see these bottles appear in big cities where the tap water tasted like chlorine for decades. But suddenly, bottled water is the norm, city and country alike.

Recently I was in Missoula, Montana, a place I hadn't been in 25 years. Back then Missoula was a small town surrounded by wild country, known as the "gateway to the Rocky Mountains." Now with well over a hundred thousand people, it is surrounded by middle-class McMansions: big sprawling houses with big sprawling lots, sprawling over the shrinking valleys and hills. Commercial development is concentrated in and around town, while agricultural activities cover much of the remaining landscape. Only the federally owned mountains in the distance remain undeveloped, though there are plenty of roads through them as well, and plenty of visitors doing plenty of things. And the grocery stores in Missoula have aisles full of drinking water, numerous brands and grades for quenching the thirst of everyone from carpenters to CEOs.

If people in Missoula, Montana, have to drink bottled water to feel safe—or simply to avoid a bad taste in their mouth—what does that say about the grandkids' water supply over the vast areas of the United States that will be far more developed than Missoula?

When you buy bottled water, you have choices among spring water, distilled water and filtered water. The spring water, of course, tastes better (if it truly comes from a spring) and is more expensive. No one should take a spring for granted. It doesn't just bubble up like upside-down manna from heaven. A spring is a natural seep where the water table, or aquifer, meets the surface of the land. Sometimes the water trickles down to a stream or brook, but most of the time it just seeps back into the ground a few feet away. In any case, a spring is a wonder to behold. Tall trees grow there and wild animals gather to drink. In dry country, you can spot a spring from miles away. All who have lived in the desert know how the sight of a distant spring brings a palpable sense of relief on a hot, dry day.

But springs can run dry, especially when you pump them. When I worked for the San Carlos Apache Tribe (which occupies a 1.8 million-acre reservation in Arizona) in the 1980s, business consultants convinced the tribe to sell bottled water from a large spring at the base of the Natanes Plateau. The plateau is the site of one of the most southwestern ponderosa pine forests in North America. Deer, turkeys, mountain lions, bears and the biggest elk in North America live in this forest. At its southern edge, the plateau ends abruptly at the thousand-foot Nantac Rim, which is inhabited by Rocky Mountain bighorn sheep. At the base are more deer, plus pronghorn antelope and javelina.

Arizona has a monsoon climate. It doesn't rain a lot in Arizona, but when it does, it *rains*. When it rains on the Natanes Plateau, which is tilted to the north, most of the water goes charging into the Black River, and much of the rest evaporates quickly. What remains seeps into the soil, providing water for the forest and its wildlife. Some of it even seeps out the bottom of the Nantac Rim, providing water for the bighorn and javelina—and now, apparently, for the water-bottling company. I asked the consultants if they knew anything about the water capacity of the plateau, and they admitted they knew nothing of the sort. But of course the thought of this 200-square-mile plateau running dry left them incredulous.

The Natanes Plateau might not go dry for a long time, but that's the point: we don't know. All we know for sure is that water demands are increasing, and the water supply is not. And the plateau is a metaphor for society's nonchalance toward water supplies. The grandkids will be even more incredulous than the water bottlers when the price for a bottle of spring water goes from $1 to $2, then $5 or more, as increasing demand ensures. And the grandkids of the San Carlos Apaches will be just as incredulous when the invisible hand of the market starts pumping the Natanes Plateau faster, when the ponderosa pines begin to thin, and when the world's biggest elk retreat across the Black River, off the reservation, heading for the White Mountains.

Of course, once spring water is exorbitantly priced, people may simply resort to the substitute of distilled water (whereupon the price of that will rise) or even, heaven forbid, tap water! We can count these as two notches—from spring water to distilled water, from distilled water to tap water—out of the quality of life for the grandkids, and these are not small notches. If you've ever quenched your thirst with a good, cold drink of spring or well water, you know what I mean.

Oh yes, and there is the fact that much of the bottled water we buy is nothing more than tap water to begin with! But that's another story.[2]

Missoula and San Carlos are among my first-hand observations related to the water supply of the United States, but most people who work with natural resources have their own water stories. Meanwhile moms and dads and even older kids, no matter how removed from the outdoors, have seen bottled water prices creeping upward. Anyone who's still complacent about water should read *Unquenchable: America's Water Crisis and What To Do About It*.[3] Robert Glennon, one of America's leading water supply experts, documents how aquifers—big ones—are running dry in the United States. Many or most regions in the world have water problems that are more dire than in the United States, most notably the Middle East, central Asia, most of Africa and much of Australia.[4] The problem isn't only water shortage; the human economy is polluting our water supplies world-wide even as they decline in quantity.[5]

If you aren't ready to acknowledge that water shortage and water pollution are real and serious problems, you should probably stop reading now. Unless you want to consider your grandkids' food.

When you go to a grocery store in the United States today, it's hard to imagine that food could ever be a problem. If you've tried growing your own food, you realize that the bounty in the grocery store is truly breathtaking! The cereal aisle alone looks like a library. But when you look at the dozens of brands, it is also humbling to remember they are all made of just a few things: wheat, oats, corn and rice, for the most part. Then there is the meat section with its hundreds of cuts, grindings and delicacies. Almost all the beef, pork and poultry, however, was raised or fattened on wheat, oats, corn, milo and soybeans. The fish section is represented by a few dozen species, and the produce section by a few dozen fruits, vegetables and herbs. That basically does it. On we go through the grocery store, seeing this basic set of species presented in boxes, bags, bottles and cans (supplemented generously by refined sugars and a host of chemicals).

Except for some of the chemicals, this bounty is ultimately dependent on three things: soil, water and sunlight. Soil and water, at least, deserve our immediate attention.

We have already considered water, but now let us tie it to food production. The fact that we get so much of our drinking water shipped to us from remote places like the Nantac Rim is partly because so much groundwater closer to town is drawn for crop irrigation. Irrigation accounts for about 40 percent of all freshwater withdrawals in the United States.[6] Our cities tend to be in plains and valleys near gently sloped agricultural areas, while the best bottled water comes from the steeper hills and mountains of the United States, Canada, Europe and Latin America. In California, where the vegetable crop alone is worth billions of dollars annually, agriculture accounts for 85 percent of water use. We are competing with our farmers, who keep most Americans and much of the world fed, for water! If this trend is not halted, at some point we will be faced with a choice between hunger and decent drinking water. Long before such a dire dilemma, of course, the city fountains will be shut off, our lawns will dry up and we won't be taking many baths.

It makes you wonder: shouldn't we cut down on some of the fountains, lawns and baths now? Some of them, at least, to buy some time for the grandkids? To buy some time while we figure out the bigger picture?

Meanwhile, the average citizen of the Western world seldom thinks about the soil, or "dirt." It's been a long time since an

FIGURE 1.1. Satellite photography of pivot irrigation on roughly 720 square miles near Garden City, Kansas. Liquidation of the Oglala aquifer sets up one of many supply shocks awaiting future generations. Credit: NASA Earth Observatory

American president warned, "A nation that destroys its soil destroys itself."[7] Soil amounts to only a few inches or, on richer lands, a few feet of the Earth's surface. When we farm, soil is exposed and runs off into rivers and eventually the oceans. Eroded soil is replaced over geological time by the decomposition of rock and organic materials, but the rate of replacement doesn't nearly keep pace with the rate of erosion. Soil erosion in the United States is ten times faster than the natural replenishment rate; for China and India it's 30 to 40 times faster.[8] It's not a declining problem, either, not even in the US where great pride is taken in the pace of agricultural innovation and technology. In the 1980s the soil lost on American farmland amounted to 1.7 billion tons annually.[9] Two decades later the figure was 3 billion tons annually.[10] It's not surprising that crop yields have been reduced over vast areas of the United States. In many areas agricultural production would be non-existent—certainly not competitive in the market—were it not for massive applications of fertilizers.

The next logical thing to consider, then, is where the fertilizer comes from, and how it gets to the fields. It comes primarily from natural gas and phosphate rock, and it gets to the fields via train, truck and tractor. The cost of phosphate rock is increasing, even faster than the price of gasoline.[11] Even for those economists who simplistically define scarcity as rising price (as opposed to an obviously diminishing resource) phosphate is becoming scarcer. Of course, for the rest of us, scarcity is a matter of common sense. A limited thing becomes scarcer as we use it up! For us, not only phosphates are becoming scarcer, but petroleum too, whether or not prices are proving it at any particular moment. Meanwhile the trains, trucks and tractors used to transport phosphates run on petroleum.

Not too long ago an economist absurdly remarked, "Worldwide, oil has been growing more plentiful, but for all we know it may some day become more scarce".[12] This telling observation was based on the fact that the *price* of oil had declined over the previous two decades. In other words, new oil discoveries of existing oil and the

FIGURE 1.2. Pivot irrigation in the Wadi As-Sirhan Basin of Saudi Arabia, February 21, 2012. Fields in active use appear darker, fallow fields are lighter. Most are approximately one kilometer in diameter. As in Kansas, the water is pumped from underground. Credit: NASA Earth Observatory

development of extractive technologies more than kept pace with demand during that period of time. But really, "growing more plentiful?" Most people know that oil is the product of organic material decay, but that doesn't mean oil is constantly being produced, making it a renewable resource like timber or fish. Oil deposits represent organic material decomposed millions of years ago in rare events that produced "source rocks," which *then* had to be buried between 7,500 and 15,000 feet below the Earth's surface to generate oil.[13] A phrase such as "growing more plentiful" is a huge red flag waving over the field of economics. It is hard to think of a good analogy for such a statement, but it is roughly akin to saying, "The food on my plate grows more plentiful, even as I eat! After all, each movement of the fork to my mouth costs me no more calories than the preceding movement. In fact, with the calories I've just converted, I'm finding it easier to move the fork, so there must be more food there, not less."

The grandkids' plight may or may not hinge upon an oil shortage. If the renowned petroleum geologist Kenneth Deffeyes is right, however, it may be *you and I* who deal with the shock of severe oil shortages. Deffeyes studied under Marion King Hubbert, who in 1956 predicted the peak of American oil production would occur in the early 1970s. Hubbert was subjected to widespread ridicule in academic and industry circles, but he was right. In fact, his prediction was a smidgen conservative, for by 1970 American production of crude oil started to fall. Three years later, the Organization of Petroleum Exporting Countries (OPEC) capitalized on this development, shocking the world with 300–400 percent increases in crude oil prices and plunging the United States and Europe into the biggest recession since the Great Depression.

Deffeyes grew up in the oil fields and spent his life in the oil business, progressing from a roughneck to a researcher of the highest scientific credentials. He built upon Hubbert's model and extrapolated it to the world, reporting his findings in *Hubbert's Peak* (2001). He predicted the peak in world oil production would occur between 2004 and 2008, giving us precious little time to develop

FIGURE 1.3. Tar sands mining operations north of Fort McMurray, Alberta, Canada. Credit: George Wuerthner

the long-promised alternatives to oil quickly enough to avoid a major depression. By October 2007, oil prices were pushing $100 a barrel for the first time in history. Prices stabilized for a while, then surged again several times over the next few years, seldom dropping as far as they rose. There will be spikes, valleys and plateaus, but we all know oil prices are never going back to those Happy Days levels again. And today there is no OPEC to blame, cajole or threaten into lowering prices and producing more oil.

There aren't any scenarios stemming from Deffeyes's prediction that aren't at least somewhat scary. A particularly scary one, however, is called the "Olduvai Theory of Energy Production." If you can imagine the topography of the Olduvai Gorge in northern Tanzania, you'll get the picture. Approaching the gorge you climb and climb in a gradual fashion, and then you break over a pleasant little ridge. Unfortunately, the downhill walk is much shorter and decidedly less pleasant, because suddenly you plunge into the gorge.

Many Americans vaguely remember the disruption caused by the California rolling blackouts of January 2001 and other localized or temporary power shortages. Hundreds of thousands in the Washington, DC region (myself included) even experienced life without electricity for days during a record-shattering July 2012 heat wave. Yet in almost any scenario following the peak of world oil production, or "Peak Oil," those old rolling blackouts and local outages will seem like child's play. Instead, we will almost surely face extended grid blackouts, as well as general breakdowns in transportation systems and other aspects of our economic infrastructure, creating havoc in the cities reminiscent of the Los Angeles riots of 1992. How could we logically conclude otherwise? The shutting down of metropolitan and regional power grids for days and weeks is unlikely to result in pleasant vacation days and caroling in the streets.

The panic meted out by the August 14, 2003, grid blackout in the northeastern United States and southeastern Canada didn't get much coverage by the media. By focusing instead on how helpful New Yorkers were to one another, the media were being politically correct with their coverage of the first major crisis in the city since 9/11. There were many helpful New Yorkers, for sure, just as there were on 9/11. But let us not fool ourselves: panic did appear in places like subways and skyscrapers. And panic is only the first problem to strike in a grid blackout. August 14 was a hot summer day. People can handle the hot weather for a day or two when water, soda and block ice are available at the corner convenience store. Things start getting ugly when the ice melts, uglier still when the soda runs out, and desperate when drinking water runs low. This is common sense, and if any verification is needed, it was provided by the nightmare in New Orleans following Hurricane Katrina. Yet for all the ugliness of an extended grid blackout in the heat of a New York summer, it pales in comparison to what could become of an extended blackout in the dead of a New York *winter*.

No one can predict precisely what a truly "Olduvaic" scenario might be, but it doesn't take a paranoid mind to conjure up images of *Mad Max* or *Escape from New York City*. There may not be any superheroes fighting off the bad guys, but there will be plenty of fighting, and plenty of bad.

Some countries will handle grid blackouts better than others, especially in summer. On July 30, 2012, a blackout in India took the electricity away from 670 million people—roughly ten percent of the global population—with outages running two days in most cases. Plenty of misery ensued, especially in the big cities, yet few deaths were reported. That's because Indians haven't been overly air-conditioned for decades. They can "take the heat" and avoid a complete meltdown in the streets.

But two days is a long way from four, eight or twenty days. The goodwill of anyone on Earth would be severely challenged by multiple weeks in scorching heat, surrounded by heat-trapping concrete, food spoiling, water running low, desperation accelerating. Goodwill would be challenged by multiple weeks of bone-chilling cold, too. Nothing good can be said about a long-running blackout except for the (unintended) saving of energy, yet the economic "analysts" and journalists miss the point every time. Regarding the biggest outage in history, the *New York Times* reported, "India's problem generating enough power has been one of the biggest handicaps to its prospects for sustaining rapid economic growth."[14] They failed to add, "India's problem has been caused by rapid economic growth, one of the biggest handicaps to its prospects for sustaining electrical power."

Ironically, the grandkids will wish we had shut the lights and fans and air conditioners off long before the blackouts, because an even bigger problem for them will be our use *of* the fossil fuels that feed our power grids. By now we all know the basics of climate change: combustion of fossil fuels releases carbon dioxide, the foremost of the greenhouse gases that trap heat in the Earth's atmosphere. In recent years the US National Climatic Data Center has made a habit of announcing that the previous year was the warmest year on record. By July 2012 there had been *328 consecutive months* with a global average temperature above the 20th-century average, indicating not an anomaly but a trend. The Nobel Prize-winning Intergovernmental Panel on Climate Change and the rest of the international scientific community has reached consensus that global warming has been a real phenomenon for decades, continues today and is accelerating with fossil fuel combustion.

Another important greenhouse gas is methane, produced by the breakdown of plant material by anaerobic bacteria. Methane is not nearly as ubiquitous as carbon dioxide, but is a far more powerful greenhouse gas. And what have been identified as the leading and increasing sources of methane? The belching of 1.4 billion cattle and the growing of 400 million acres of rice,[15] which brings us back to the meat and cereal aisles, respectively. Meanwhile, the breakdown of chemical fertilizers (depended upon to produce the cereals and meats) puts at least seven million tons of nitrous oxide, another greenhouse gas, into our atmosphere.[16]

Doesn't it seem like something is truly, horribly awry? If we keep stocking the grocery shelves with an increasing volume of grains, meats and vegetables via fossil fuel combustion and fertilization, we court soil erosion and global warming. If we stop the fertilization and the fossil fuel combustion, we pay more for our food and start going hungrier. Either way we forego good drinking water, in the first case by using more water for agriculture, and in the second by sacrificing the spring water trucked in from afar. But of course this is not really a choice. Hunger and thirst are powerful, primal motivators. The next-to-last thing we will do is go hungry, and the last thing we will do is go thirsty. People, tribes and nations will fight for food, as they've done repeatedly in sub-Saharan Africa, and they will fight even more stridently for water, as they've done in the Nile Basin and the Middle East.[17] Hunger and thirst, in other words, may be the last things experienced by many of the grandkids as their agricultural and transportation systems break down. Dying of hunger or thirst—or living with the crippling effects of malnutrition or dehydration—is not a good way to go.

Only the muddle-headed would call this a misguided "Malthusian" analysis. When agricultural and transportation systems break down, it is precisely hunger and thirst that transpire, especially with so many people dependent upon bottled, transported water.

Meanwhile, according to economic theory and history, fossil fuel combustion will not be abandoned as long as the price of fossil fuels *does* remain competitive, for example if Peak Oil turns out to be more of a mesa than an Olduvai Gorge. There is little incentive for corporations, who conduct most of the research and development in the United States, to develop alternative methods. Most of the rest of American research and development is conducted or funded by the government, which has been shockingly slow to respond. President George W. Bush balked at even *acknowledging* global warming, much less planning to reduce the rate of it. The American government under President Barack Obama has done plenty of acknowledging and tidbits of planning, but no accomplishing. If ample alternatives for powering our agriculture and transportation do exist, we are way behind in developing them.

Because global warming is causing the volume of ocean waters to expand and glaciers to melt in the mountains, at the poles and over the Greenland ice shield, sea levels are rising.[18] Approximately half of all Americans live within 50 miles of saltwater coastline and the proportion is growing. Substantial areas of these coastlines will simply be submersed, beginning of course with low areas in Florida, Louisiana and Maryland. The problem isn't limited to populous areas, either. Vast acreages of our saltwater marshes, coastal forests and other tidal habitats have been, are being and will be inundated and lost. Louisiana alone loses an average of 16.6 square miles every year. That's a football field an hour of some of the most valuable areas in the world for fish and shellfish production.[19] Not all of Louisiana's coastal problem is caused by sea-level rise—the other reasons are much more directly related to the economy[20]—but sea-level rise is a significant factor. Meanwhile freshwater aquifers near the coasts are being inundated with saltwater, putting additional pressure on inland aquifers for water production. In other, steeper areas, such as the California coast, sea-level rise helps to usher the terra firma out to sea in dramatic and unpredictable fashion.

In other words, the increasing amount of food on our shelves, supported by the increasing combustion of petroleum and application of chemical fertilizers, sets up a population displacement program of overwhelming proportions. This is sound logic that squares with common sense. The grandkids will have to return to areas their grandparents found less desirable, and those areas will be far more crowded and congested this time around. Industry too will be seeking higher ground, and employees will have to adjust their movements likewise. Coastal disruption will be accompanied by increasing pressure on inland infrastructure such as roads and utilities. The fact that this general trend will happen is indisputable because it's already begun.[21] The only real questions are how fast it will happen, to what degree, and how fast we can adapt. Widespread hardship is almost certain, and chaos is not out of the question.

My focus has been on the US, but American problems will probably pale in comparison to the problems in sweeping lowlands from Rotterdam to Bangladesh. Demand for American agriculture will reach alarming proportions, and countries populated by the hungry will be just as motivated as their hungry citizens. In the past, the US looked on with compassion and, in many cases, provided voluntary relief. If the relief was insufficient, there was little a starving nation could do about it. But in a world where "developing" countries like India and Pakistan are flexing their nuclear muscle and others, at least as horribly, have been honing their biological weaponry, how much security will the grandkids (here, there or anywhere) have?

Fossil fuel combustion and chemical fertilization are not the only unsustainable forces propping up agriculture. In order to provide the increasing quantities of groceries we see in the store, agriculture has been industrialized all the way from the field to the retailer. In the field, industrialization means simplification. Vast landscapes in the United States are now devoted to one type of crop. Crops have also become simplified genetically to make them grow identically, making it more efficient to harvest and process them with evermore specialized equipment.

This is not how Mother Nature rolled. Her landscapes were chock full of variety. Even where single species dominated large areas, such as big bluestem on the American tallgrass prairies, they were constantly challenged by other plant species growing in their shadows, all of which were perpetually subject to insect predation and disease. The result was a rich display of physical and genetic variety which served as survival insurance against specialized competitors and predators. Mother Nature rolled in all kinds of weather, all kinds of astronomical anomalies, all kinds of volcanic activity, all acts of God.

Today's monocultures, by contrast, are highly susceptible to blights, parasites, predators and weeds, which are then battled with fungicides, herbicides and other pesticides—"agricide," as the war has been called.[22] As early as the mid-1980s more than 3,000 pesticides had been used in the United States, despite the fact that toxicity data were lacking or insufficient for most of them. Today, over two billion pounds of more than 18,000 registered pesticides are used every year in the US alone.[23] Do we think the government is keeping up with the daunting task of monitoring the effects of this burgeoning array of chemicals on our health and our ecosystems? I don't know about you, but I have my doubts. Worldwide, pesticide expenditures are nearly $34 billion annually, about a third of which is spent in the United States.

The most recent stage of agricide, however, is the genetic modification of crops. Agricultural monopolies like Monsanto and jealous rival conglomerates like Dupont are manipulating the genes of crops to "produce" pesticides on the one hand and to withstand pesticides on the other.[24] For example, "Bt corn" is engineered with a gene from a bacterium, *Bacillus thuringiensis*, that produces a chemical toxic to corn borers, a major corn pest. So the pesticide is built into the corn in one neat package. There's no rubbing it off on your shirt, like dinocap (a fungicide) from an apple. Meanwhile, Roundup Ready soybeans are engineered to withstand glyphosate, the active ingredient in the herbicide Roundup. This method of genetic engineering leads to increased usage of pesticides, but at least the pesticides aren't built into the crop itself.

As they say, "pick your poison."

Then there are the antibiotics, growth stimulants, pigmentation enhancers, appetizers and other chemical additives used by the beef, pork, poultry and fish industries. By the time a cut of beef gets onto the grocery floor, the steer has been exposed to a chemical environment from the milk he suckles to the fat he puts on prior to slaughter. Many of these chemicals become concentrated in fatty tissues in a process called biomagnification. Then, we eat them. No wonder a growing cohort in the medical community recommends organic foods for the prevention of arthritis, liver disease, hyperactivity and a host of other maladies.

But were agribusiness simply to abandon these chemicals across the board, grocery prices would rocket through the roof and hunger would ravish the poor. Do you doubt it? Most readers would be financially stressed to subsist entirely from the organic section of the grocery store. Now imagine the billions of people under the poverty line trying to make a go of it from the organic section. And we're only in the early stages of Supply Shock. The fact is that the vast majority of us have little choice but to ingest our share of these chemicals, many of which are proven or suspected carcinogens (and some of which are suspected mutagens). Is it any wonder that the rate of cancer in the United States and Europe has been steadily increasing? What will the rate be for the grandkids? There goes another slice of life.

Thus far we have only explored one economic sector (that is, agriculture), a sector that accounts for about 16 percent of American expenditures. The manufacturing trades are too numerous to survey here. But one example serves to illustrate what lies ahead for the grandkids as a result of the manufacturing economy.

The refrigeration industry was a consumer health concern as far back as the 1920s, when dangerous chemicals like ammonia and sulfur dioxide were the primary refrigerants. In 1928 a Dupont chemist discovered chlorofluorocarbons (CFCs), which were nonflammable and nontoxic. Eventually, CFCs were used as aerosol propellants in a vast array of domestic products, culminating in an $8 billion dollar industry in the United States.

In the early 1970s, following an incredibly lucky choice of research topics, a handful of chemists developed the theory that: 1) CFCs were a stable class of molecules that slowly rose to the stratosphere (that portion of the atmosphere 10–30 miles above the earth's surface); 2) CFCs in the stratosphere were broken up by the sun's ultraviolet radiation to produce chlorine atoms; and 3) the chlorine atoms initiated a chemical reaction that destroyed ozone.

An ozone molecule consists of three oxygen atoms and is highly unstable. If all the ozone in the stratosphere were compressed under the air pressure found along the Earth's surface, it would comprise a blanket about one-eighth of an inch thick. This delicate blanket, unfolded loosely in the skies, is what prevents carcinogenic levels of ultraviolet radiation from reaching the Earth's inhabitants. It is literally a shield for human survival.

At the time the ozone depletion theory was developed, almost a million tons of CFCs worldwide were being released into the atmosphere. Unfortunately for the grandkids, most of these tons will not enter the stratosphere until the latter half of the 21st century, at which time some 7 to 13 percent of the ozone will be destroyed—"enough ozone depletion to seriously alter life on earth."[25]

Science journalist Sharon Roan documented the scientific, political and economic history surrounding the ozone depletion theory in her 1989 book, *Ozone Crisis*. The refrigeration industry fought hard to discredit the theory, but as more atmospheric scientists studied the theory, they reached consensus. During the 15 years—*yes, 15 years*—it took for this consensus to defeat the CFC industry via the Montreal Protocol, millions of tons of CFCs were sent skyward to take another notch from the grandkids' future. This frustrating deadlock was enough to make innocent people cry, and in some cases fight, not that the old "captains" of the CFC industry were anywhere in reach.

And here we have one out of tens of thousands of chemicals that have gone into production for the sake of economic "efficiency," thus contributing to economic growth. Many of these chemicals are seemingly harmless when discovered, but let us invoke our common sense: How can we possibly know the ecological ramifications of a chemical, especially one foreign to Earth's natural environment? What do we begin to test for? Knowledge of flammability and toxicity, for example, was worthless to the CFC debate. From the asbestos in our attics to the saccharine in our sodas, people of the industrialized nations have witnessed a history of miracle compounds that turned out—belatedly—to be hazardous, sometimes with heart-rending persistence. Yet the burgeoning list of chemicals is touted by economists as a symbol of the "technological progress" that facilitates economic growth.

As for the grandkids, the question is: How many of these miracle compounds have we failed to discover the ill effects of, while we produce more such compounds by the score? As the American tobacco industry has so amply demonstrated, industry cannot be counted on to reveal the hazards of their products. If anything, industry may be expected to *conceal* such hazards. ("Nicotine is not addictive," you may recall.[26]) As with CFCs, theories outlining the hazards of a product will be portrayed as "preposterous" until the scientific consensus becomes overwhelming. But when you dig deeply into your common sense, would it be at all preposterous to think there are some nasty surprises awaiting the grandkids? And that the number of such surprises will increase proportionately to the number of synthetic compounds we develop for the sake of economic growth?

This, I think, is enough. After all, if we're paying any attention, we are exposed to these types of issues day in and day out via newspaper, public radio, television, the Internet and our own personal experience. It is important to realize before we go further, however, that we have been talking only about *one sector* of the economy (agriculture), plus *one example* from the manufacturing sectors and some observations about the energy sector that keeps it all humming. We have not begun to look at the extractive sectors like logging, mining and fishing, much less the vast sweep of manufacturing sectors, ranging from the heaviest (such as metal ore refining) to the lightest (computer chip manufacturing). Nor have we looked at the infrastructure and service sectors needed to keep it all afloat. Nor the indirect toll on the quality of life wrought by crowded conditions, species extinctions or perpetually diminishing wild country. And lest we forget, when the grandkids face these horrendous challenges, they will do so in an increasingly congested, noisy and dangerous environment.

The main reason for stopping here, however, is that the knowledge *of* these problems is good for only one thing: developing the conviction to address them. For those without that conviction, there is no need to belabor the point. Besides, plenty of books are available to describe the impending environmental and socioeconomic crises in far greater detail. The Worldwatch Institute's *State of the World*, for example, has been doing this annually for decades. The task ahead is to develop the knowledge of *how* to address the situation.

For each individual problem, of course, there seems to be a clear technological fix. Stop burning fossil fuels and emitting CFCs. Stop applying chemical fertilizers and pesticides. Stop drilling wells and tapping springs. Figure out another way here, another way there. Soon enough, however, the list becomes overwhelming, and one does not know where to turn next. And all of these actions cause prices to rise. Meanwhile, a new threat seems to loom every day. With the situation appearing so mind-boggling and hopeless, it is easy to see why people give up, hoping some technological breakthrough will come along to save the grandkids. Some are probably praying that, at the final hour, humanity can escape into outer space.[27]

But it is not so hopelessly complex. There is *one simple process* driving all these problems: economic growth. Halting economic growth now will not guarantee a healthy, happy future for the grandkids, but it will at least allow for one. The first prerequisite is slowing, then halting economic growth in the United States, Europe, Japan and other highly developed, wealthy nations. That includes halting the desperate measures to stimulate the economy at the obvious detriment to the environment and to fiscal stability. Abandoning the growth path entails some sacrifices, most likely including higher prices during the adjustment phase. The sooner economic growth is halted, however, the better the prospects will be for the grandkids to establish an acceptably-sized global economy. After these grandkids struggle through the adjustment phase, succeeding generations may be positioned for environmental health and stabilized prosperity.

Economic growth *can* be halted before we breach the absolute limits imposed by Mother Nature. It will take leadership, entrepreneurship and engineering, but this time applied to public policy and not to mere "stuff." It will not be easy, but it will be easier than tackling each of the aforementioned, unmentioned, and as-yet unidentified crises one by one. It is also more politically feasible to work at the macroeconomic level, as we will find in Chapter 11. Economic growth can be halted especially in the democracies of the world, even "capitalist" democracies (which are really mixed economies) such as the US. While capitalism is not particularly conducive to the halting of economic growth, democracy is. The first principle of democracy is majority rule. Once the majority of citizens come to understand why economic growth has become such a threat to their grandkids, the democratic process can help us with how to alleviate this threat.

CHAPTER 2

Good Growing Gone Bad

> *Growth in GNP is so favored by economists that they call it "economic" growth, thus ruling out by terminological baptism the very possibility of "uneconomic" growth in GNP.*
>
> HERMAN DALY

CHAPTER 1 IMPLICATES economic growth as the cause of major environmental problems, even disasters, so now it is important to consider precisely what economic growth is and how it occurs. Fortunately, the subject is not as mysterious as implied in titles such as *The Mystery of Economic Growth* or *The Elusive Quest for Growth*. People like to read mysteries, and economists like to sell books, but economic growth is just an increase in the production and consumption of goods and services in the aggregate.

Goods and services, on the other hand, have had their mysterious moments. Certainly the distinction between goods and services has mystified many, misleading them into fallacious and dangerous arguments. Most notably, some believe in the notion of a "service economy" growing evermore independent of goods. This is called the "self-sufficient services fallacy,"[1] and will be further described (and debunked) in Chapter 7, but here we stick with the basics.

Goods include all the physical objects you can purchase, use or consume. Boots, automobiles and carrots are goods. If it's a good, you can place it on a scale and weigh it. Services include everything else you can buy or that benefit you. Entertainment, retirement insurance and window-washing are services. You can't put them on a scale and weigh them, but they all entail physical items and movements, many of which can be weighed (like the professional dancer) or otherwise measured (like calories burned by the dancer).

Production and consumption are two sides of the same coin. Generally speaking, what is produced is consumed. In fact, this is such a reasonable principle that it became one of the earliest laws of economics. It was called "Say's Law" after Jean Baptiste Say, whose *Treatise on Political Economy* was published in 1803. A lot of cold water has been thrown on Say's Law, especially since the Great Depression when a glut of goods and services was produced but never purchased by consumers. Many economists claim Say was wrong, but they tend to forget that Say wrote at a time when bartering was still a prominent form of exchange outside the larger cities.[2] In a bartering economy, with people trading goods and services directly, one cannot "buy" (consume) unless one "sells" (produces). If the farmer wanted to obtain a scythe, he had to produce something to barter with—say, 20 bushels of wheat. The loop was closed, heads came with tails; production equaled consumption.

Money, on the other hand, can sit in banks and complicate matters. Credit complicates matters even more. We know that some goods are produced and go to waste rather than being consumed. Fish can sit in the barrel too long, chukka beads go out of style, and iPads go obsolete as we speak. Unless these items are quickly marketed at lower prices, perhaps much lower prices, they will go to waste. Nevertheless, a lot can be said for Say's Law. It is weak as a law but strong as a rule of thumb, even in a modern economy with cash, checking and credit accounts. Producers seldom produce unless their products are quite likely to be consumed. Why? Because when products aren't purchased, producers lose money. The "invisible hand" of the market swipes the unwanted goods and services away, producers learn quickly and Say's Law is generally kept intact.

Governments keep track of production and consumption with a method called national income accounting. In the United States, national income accounting has been conducted since 1929, mostly by economists in the Bureau of Economic Analysis. The work of these staid accountants, as in most countries, is based upon the fundamental identity of national income accounting. (In mathematics, an "identity" is a type of definition.) The fundamental identity of national income accounting states that total production equals total expenditure equals total income. In other words, by definition, the monetary value of goods and services produced in an economy will equal the amount of money spent, which in turn will equal the amount of money received.

If you think that sounds like a slightly fancier version of Say's law, you're right, and the Bureau of Economic Analysis has developed an elaborate approach to ensure the three categories (production, expenditure and income) amount to the same each year. It's not that they're cooking the books, either. Rather, they know that in the real world, income matches production and requires expenditure, so their accounting is structured to fit that reality. Using the production approach to national income accounting, economists estimate the amount of output (goods and services) produced. More specifically, they calculate the market value of final goods and services produced during a year.[3]

The expenditure approach entails estimating the total spending on final goods and services during the year. Household consumption is typically the biggest category of spending in a national economy. In the United States, for example, it accounts for about two thirds of spending. Investment, government purchases and net exports are the other major spending categories.

The income approach identifies five primary recipients of income: employees, landlords, small businessmen, corporations and lenders. For employees, the main categories of income are wages, salaries and other benefits. Landlords receive rental income. Small businessmen and corporations receive profits and lenders receive interest. Adding them up produces a number called national income. When depreciation and a few other things are accounted for, the resulting figure will be identical to the production and expenditure tallies.

The amount of goods and services produced within the boundaries of a nation is called "gross domestic product," or GDP. Each approach to national income accounting—production, expenditure and income—produces an identical GDP figure. One may also account for the goods and services produced by domestically-owned firms regardless of where the production takes place, in which case the size of the economy is referred to as gross national product, or GNP. For example, Japanese firms produce many of their goods and services within the boundaries of other nations. Therefore, Japan's GNP is considerably higher than its GDP.

GNP was once the standard measure of national economies. However, GDP is simpler to understand and is more clearly relevant to what happens within a country's borders, so it has eclipsed GNP as the standard. Politics were involved as well. In the US, the shift toward GDP as the measurement of choice occurred during the administration of George H.W. Bush. Leading up to the 1992 election, in a country prioritizing economic growth, Bush needed economic bragging rights to have a chance against Bill Clinton. With the firms of Japanese and other nations growing faster in the US than American firms growing outside the US, Bush was able to claim a higher rate of growth by using GDP rather than GNP.

Meanwhile, the Japanese still prefer to report their economic exploits in terms of GNP. Japan has a higher GNP than all countries except the US and China,[4] but in terms of GDP, the European Union and India are also ahead of Japan.[5]

Regardless of politics, GDP is a fine measure of economic activity within a nation's borders. GDP makes it easy to compare the sizes of economies around the world and over time. For example, in 1990 US GDP was approximately $8 trillion, while Canada's GDP was approximately $800 billion and Colombia's was approximately $80 billion.[6] In other words, based on GDP, the American economy was ten times as large as the Canadian economy and one hundred times as large as the Colombian economy.

Economic *growth* occurs when *more* goods and services are produced, more money is spent and more income is received from one year to another. For example, American GDP grew to approximately $12.8 trillion in 2001, an increase of $4.8 trillion since 1990.

These figures were reported in "constant dollars" or "real dollars," meaning they were adjusted for inflation.[7] Using real dollars allows us to compare the real effects of economic growth on purchasing power. Real GDP figures are extremely useful indicators of the material size of an economy, much like grams and ounces are useful indicators of material size, whether they are measuring the weight of a diamond or a drop of water. Some economists may not like this analogy, but they haven't read Chapter 7 yet. Meanwhile, common sense enables most of us to see how a country with a large GDP tends to have a larger material presence than a country with a small GDP.

One of the most dramatic cases of economic growth in world history is now taking place in China, where the primary currency is the yuan. In 2012, an American dollar was worth about 6.5 yuan — you'd get about 6.5 yuan for a dollar — but it doesn't matter when thinking about how fast the Chinese economy has *grown*. Chinese GDP (adjusted for inflation) grew from approximately 2.4 trillion yuan in 1992 to 17.7 trillion yuan in 2012.[8] In other words, the Chinese economy is more than 7 times bigger than it was 20 years ago when the first McDonald's restaurant opened. Since then perhaps

FIGURE 2.1. Energy and agricultural sectors side by side at Xi'an, capital of Shaanxi province, China. Credit: David Klotz

the most symbolic episode of Chinese economic growth has been the construction of Three Gorges Dam, the world's largest power station, which displaced well over a million people, obliterated an ecosystem's worth of wildlife and gave a final water burial to a thousand archeological sites.

India is the other huge, fast-track economy. Indian GDP grew from approximately 15.7 trillion rupees to 59.6 trillion rupees—four times bigger—in the same two decades.[9] India isn't known for resource-extraction dramas such as the damming of Three Gorges, the strip-mining of tar sands in Alberta or the mountain-top mining of Appalachia, but it is well known for out-of-control pollution problems, punctuated by the most poisonous of all, the Bhopal disaster of 1984. That's when a Union Carbide gas leak killed thousands of Indians and exposed hundreds of thousands of others, plaguing victims for decades with respiratory problems, eye diseases, neurological disorders, cardiac failure and birth defects. Today, Bhopal is largely forgotten outside the state of Madhya Pradesh, but India is known for the cumulative effects of 1.2 billion souls making a living, or trying to, in a country one-third the size of China.

Not that there is a Roper poll proving that India is known for the cumulative effects of getting on with life, but who would like to be randomly plopped, mouth open for more than an instant, into a river in India? Not only rivers near the iconic landfill sites around Mumbai or New Delhi, but any rivers in a country of 1.2 billion?

It is instructive, when considering the size of the global economy, to remember that the only reason the Mumbai landfill images are more iconic than those of, say, Seoul, Sao Paulo or Los Angeles is the photos of kids climbing them. Probably only the Olusosun landfill of Lagos, Nigeria, would rival the Deonar landfill of Mumbai for the title of most photographed children. But whether kids are navigating them or not, landfills are the end of the tailpipe of economic growth, and landfills dot the planet in growing numbers and growing sizes. In the US alone, 250 million tons of trash are tossed every year.[10]

It's worth imagining the footprint of 250 million tons for a moment. Smaller cars in the US weigh a ton. Imagine 250 *million* smelly garbage piles that each weigh what a car does. What happens to all this annual garbage? Of course it goes into the growing landfills, but then it percolates downstream. You wouldn't want to be plopped into many American rivers, either.

It may seem self-evident, but it warrants emphasis that two major forces are at play in the process of economic growth. One is the number of consumers, and the other is the amount consumed by each consumer. Here we are using the word "consumer" in the broadest sense: households, firms and governments. We also use the word "consume" in *its* broadest sense: current consumption and investment. For the economy as a whole, consumption is indicated by expenditure. The average consumer of a nation is therefore described by the total amount of expenditure (GDP) divided by the number of citizens. From here on, we'll refer to the consumption of this average consumer as "per capita consumption" or "per capita GDP." Given the fundamental identity of national income accounting, per capita consumption equals per capita income equals per capita production.

The number of consumers in a nation is the same thing as its population, because everyone in a population must either consume or die. If a nation's population grows and per capita consumption stays the same, the nation's economy grows at the same rate as the population. A doubling of the population, for example, results in a doubling of the size of the economy. Likewise, if population remains the same but per capita consumption rises, the economy grows at the same rate as per capita consumption. Doubling per capita consumption, for example, results in a doubling of the economy's size. When population and per capita consumption increase simultaneously, the effect on GDP is multiplied. If population and per capita consumption each double, the economy grows to four times the size of the original.

The contribution of government expenditures and corporate investment to GDP has to be considered when we compare per capita consumption among nations. For example, if a nation's GDP consisted entirely of household goods and services, then per capita consumption would be an excellent measure of the amounts of goods and services actually consumed by average folks. Invariably, though, a significant share of a nation's GDP consists of government expenditures and the capital investments of corporations, so that per capita consumption doesn't necessarily provide an accurate measure of household economic conditions.

Of course, many government and corporate expenditures may ultimately contribute to the economic welfare of citizens, because governments and corporations employ citizens who then spend their wages on household goods and services. However, many government and corporate expenditures tend to distribute wealth in ways not favorable to the average citizen, as when a navy purchases a fleet of extremely expensive warships from a large shipping corporation. Large corporations are invariably and notoriously structured such that the top executives benefit from such transactions far out of proportion to the average citizens they employ, as described in *When Corporations Rule the World*[11] and, with a more institutional focus, in *The Corporation: The Pathological Pursuit of Profit and Power*.[12]

Meanwhile, GDP does not account for the various economic activities performed outside the marketplace. For example, household services performed by family members are not accounted for. In some nations, especially so-called "undeveloped" nations with less economic specialization, these household activities are well developed. Nor does GDP account for goods and services produced and consumed in the "black market," where deals are done in the dark and cannot be accurately accounted for by governments. A good example in the US and Europe is the production and consumption of illegal narcotics. Therefore, GDP is not an entirely accurate measure of the economic activities of a nation. It tends to underestimate the size or heft of an economy. A dramatic, hypothetical example of underestimating with GDP would be if a nation turned entirely to barter. GDP would decline to zero, yet all the people would be vigorously producing and consuming. Although this is an extreme hypothetical example, used simply to make a point, it is relevant because the incidence of bartering (relative to monetary transactions) for real goods and services tends to increase during financial and economic crises.

To summarize, economic growth is simply an increase in the production and consumption of goods and services in the aggregate. It results directly from increasing population, increasing per capita consumption, or both. It is facilitated by technological progress, but that is a topic for Chapter 8. Household goods and services typically account for the bulk of a nation's production and consumption unless a major war is underway. Government expenditures and capital investments tend to muddy the waters of GDP calculation, as do black markets, bartering and unpaid-for domestic services. Nevertheless, GDP is a good index or indicator of the overall economic activity occurring in a nation. When we add up the GDPs of all nations, we also have a good indicator of the economic activity on the planet—the heft of the global economy—and that activity went from approximately $30 trillion in 1990 to approximately $48 trillion in 2009.[13] As *Supply Shock* went to print in early 2013, the size of the global economy was approximately $79 trillion in 2011 US dollars, partly reflecting rapid and real growth (such as of China and India) and partly reflecting inflation.[14] No matter how we look at it, the global economy grew at a remarkable rate in the first dozen years of the 21st century.

Population trends are important to consider when interpreting GDP data. Consider what happens to national income when an increase in GDP results primarily from the immigration of wage laborers. With the influx of these new consumers (and cheaper labor), the income of businessmen, corporations, lenders and especially landlords increases. This is offset, however, by a decline in the income of the average wage laborer. Immigrants will work for lower wages than other laborers, especially at first, but wages are eventually driven down across the board. That's why labor unions often oppose liberal immigration policies. Meanwhile, despite how politically conservative Big Money or Capital is, it favors liberal immigration laws, as well as liberalized (unregulated) trade.

If you've ever been confused by the term "neoliberal," now you shouldn't be! Neoliberals are free-market, social conservatives. They can't be called plain old liberals because that term has been reserved for those who see the need for government to play an active role in economic affairs, including international trade. The irony is that American neoliberals who also promote American dominance in international affairs are then called "neoconservatives." (Remember Paul Wolfowitz?) For a neoliberal to be a neoconservative entails some tricky diplomacy.

Now consider a rising GDP with a stable population. This indicates a rising per capita income. Again, the higher incomes may not be distributed equitably among the population, but typically an increase in per capita income in a nation with a stable population is financially helpful to most people. Economists consider an increase in per capita income one of the most important measures of success, and so does the average citizen in many countries, at least at this point in history.

This quick comparison of the two major types of economic growth (one based on population, the other on per capita consumption) provides an immediate insight into the desirability of economic growth. If economic growth occurs but population growth occurs at the same rate, what is gained by the average person, and especially the average wage worker? Some win and some lose, but on average, nothing is gained unless the average person likes to be surrounded by more people. If a nation is sparsely populated, many of its inhabitants may welcome the addition of people, with or without increased per capita GDP. If the nation is sufficiently crowded, however, the addition of more people would seem to result in a *declining* quality of life, even though per capita GDP remains the same. Such would seem to be the case in nations where traffic congestion, noise pollution and generally crowded conditions are common.

As an American who has lived all over the US, I think this has clearly become the case in my home country. Urban sprawl and traffic congestion are major issues. Crime related to crowded city conditions, meanwhile, has been a constant challenge for many decades. Larger and larger areas are becoming noisy and stressful environments. In many areas of the country, one has to travel a considerable distance to find any solitude. Western European countries have long experienced similar conditions.

In fact, don't crowded conditions characterize many "undeveloped" or "developing" nations as well, or even more so? India and China, most notably, have wrestled with intractable population problems for decades. Worse yet are the ongoing or recurring tragedies in Bangladesh, Haiti and too many African countries to list. Some will claim that in many cases the problem is not a matter of population growth but rather ethnic or tribal rivalry. Common sense answers that such "rivalry" is a function of how large the rival groups grow. Population growth brings factions into conflict for scarce resources.[15]

Over vast areas of the planet, then, it seems the addition of people is simply making life more stressful, leading not only to urban sprawl but to higher rates of stress, crime and disease. The more subtle problems associated with water, agriculture and pollution discussed in Chapter 1 are also directly linked to population growth. This is the situation called the "full-world economy" by Herman Daly, the ex-World Bank economist who won a Lifetime Achievement Award from the National Council for Science and the Environment.[16]

Herman Daly is a hero of mine, a brave and brilliant iconoclast who wasn't bought out by Big Money or the "economic hit men."[17] Yet I've met plenty of Americans who don't think we are living in a full world. These people generally fall into two categories. The first consists of people who have lived their entire lives far removed from the land. In other words, they've lived exclusively in bustling cities or large suburbs, and it's easy to understand why they would have a higher tolerance for crowds. These folks tend to be sociable but feel out of place in the forests, plains or mountains. Take Woody Allen, for example, known for his love of New York City, and for acknowledging, "I hate the country."[18] Naturally, such people tend to have little knowledge of agricultural and extractive activities, and see little reason why the landscape shouldn't be one big city.

Nor do they fully grasp the importance of the agricultural and extractive sectors to city life. In rural areas, a derogatory remark about city dwellers is, "They think milk comes from a carton." Hopefully that's an exaggeration, but neither can we expect long-time city dwellers to understand how much room and how many resources the cows need to produce milk for millions of people.

The other significant category of people who don't think we live in a full world are found on the other end of the cultural spectrum. These are the folks who *have* lived in the country, but way out in the country. They tend to be loggers, miners, cowboys and ranchers. They wake up to wide-open vistas, drive on wide-open highways and fall asleep under wide-open skies. Many of them love it and many take it for granted, drifting along like tumbling tumbleweeds, like some modern-day Marty Robbins. I doubt there are many such folks living in Europe or Japan, and their numbers are quickly dwindling in the US, but they still occupy significant portions of the American West, especially in the Great Plains, the Great Basin and the Sonoran Desert. They don't realize they're a dying breed. They seldom witness the cityscape pushing its way across the landscape, and going to town is a welcome respite from the physical toil and loneliness that disturbs their peace. In other words, they are at the other end of the spectrum from the "city slickers." The view from either end of the spectrum is too limited to see what ails the spectrum as a whole.

I think most of the remaining Americans, however—those residing along most of the spectrum—are beginning to sense a problem. Midwestern farmers, New England fishermen, coastally pavemented Floridians, Pacific Northwesterners lined up for coffee, and especially the many southern Californians moving to Texas and Colorado come to mind. My sense of European politics (based upon the news and my own travels) is that a majority of Europeans are feeling uneasy as well, especially in the crowded western countries but also in the northern countries where previously vast stocks of solitude are visibly being liquidated. Clearly the Indian and Chinese governments have histories of concern with the fullness of their worlds, too. From Austria to Zimbabwe, surely "fullness" is creeping into the collective consciousness of citizens worldwide.

Virtually everyone, then, is affected by the process of economic growth, even those who are not so concerned with personal income and financial affairs. Economic growth may clearly have negative as well as positive effects, and the negative effects tend to affect all citizens, whether they agree with the goal of economic growth or not. Except perhaps for a slight minority of hermits, everyone must eventually contend with congestion, noise, and the general stress of encroachment and crowded conditions, all of which are byproducts of a full-world economy. Even the hermit's quality of life may be diminished by insidious environmental problems such as ozone depletion, global warming or water pollution. At some point, even being a hermit becomes impossible because there is no place left to hermitize.

The inevitable gut reaction of some citizens is, "If economic growth was such a good thing for so long, why is it suddenly a bad thing?" Yet suddenness should be expected to separate good growth from bad. For example, if we could all agree on a number that constitutes a full world, say 7 billion people, we see that the world goes from not full to full in the first breath of the 7 billionth baby. This happened on October 31, 2011, while I was writing this book. Alternatively, if we all agree that a gross world product (GWP) of $70 trillion constitutes a full world, we see that the world may go from not full to full in the purchase of a Chicken McNugget or some other good or service. This too happened in 2011.[19] If $70 trillion marked the full level, most of us already would have considered the world very close to full for some years.

The point is, however, that one day we awoke, or will soon awake, to a world that is full by almost *everyone's* standards. This is a sudden and rude awakening if we have not been alert to the prospect. It's far too late to rectify it with one less McNugget.

Who would you trust to estimate how much capacity our planet has to support the global economy? No idea? Well let's start with who you wouldn't trust. Common sense and experience would tell you not to trust British Petroleum, Wall Street, OPEC, the World Bank or the Tokyo Stock Exchange. Most of them couldn't be trusted to tell the truth about economic prospects, generally speaking, and those with less incentive to lie (perhaps the World Bank, for example) still wouldn't know what they were talking about. Some of the best estimates—the most scientifically grounded and least motivated by profits or loyalties—are provided by a non-profit organization called the Global Footprint Network (GFN). The GFN was established by scientists with a concern for sustainability and has become the go-to source for information on economic carrying capacity. They recently summarized, "Today humanity uses the equivalent of 1.5 planets to provide the resources we use and absorb our waste. This means it now takes the Earth one year and six months to regenerate what we use in a year."[20]

In a full world, where economic growth based upon the growth of humanity is no longer desirable, we would only encourage economic growth based upon increases in per capita consumption. Presumably this would increase our quality of life.

Or would it? We must now look more closely at the consequences of economic growth based entirely upon per capita consumption.

Those who clamor for economic growth based purely upon per capita consumption—sometimes called "affluence"—neglect a few important principles of economics. One is the law of diminishing returns, which states that, as increasing amounts of goods and services are purchased, the utility, usefulness or desirability of an additional purchase declines. For example, a house is a very useful thing. A second house is less useful, and a third may just become a burden. A play is fun to go to every month, perhaps, but who wants to go to a play every night? Every good must be maintained, and every service requires our time. At some point the maintenance costs for goods become prohibitive and the time spent enjoying ser-

FIGURE 2.2. Refining, manufacturing and transportation sectors meet along the Chesapeake Bay near Baltimore, Maryland. Chesapeake Bay was once the most productive estuarine system in North America. Credit: IAN Image Library

vices is better spent playing with children, visiting neighbors and participating in town-hall meetings.

Another principle of economics is called co-production, a concept actually rooted in chemistry. For every valuable thing produced there is something else produced that is not so valuable. The most obvious example of co-production in the economy is pollution. For every ream of paper that rolls off the factory floor, some sulfur dioxide drifts downwind. In most economic scenarios, however, the concept is not as clearly one-to-one as in simple chemical reactions. In addition to sulfur dioxide into the air, the paper company sends a stream of chloroform into the water. In fact, there is a vast smorgasbord of by-products associated with paper, plastic, film and other chemically-laden manufacturing processes.

The concept of co-production is less obvious in the production of services, but it's obvious enough when we think of the goods those services entail. The entertainment sector sounds fairly innocent of co-production, but think of the chemical soup entailed in the production of the NASCAR, Grand Prix and Daktari racing fleets, not to mention the fumes, exhaust, particles and junk generated in the racing thereof. Not fair to pick on one entertainment sector? It's fair enough when we consider that NASCAR has been the fastest-growing spectator sport in North America over the past several decades.

In a full world, all this co-production causes real problems for real people, diminishing the quality of life even as GDP climbs and long after the fun has been had. While certain types of pollution have been addressed to some extent, there is always a new breed of pollution nibbling away at our peace of mind and most likely parts of our bodies. In the developed nations, at least, diminishing returns and co-production are enough to cast serious doubt upon the desirability of further economic growth, even based on per capita consumption.

And then there's space. We saw how clearly population growth leads to a full world. In full nations and ultra-full cities such as Mumbai, Shanghai, New York, Los Angeles, Seoul, Sao Paulo, Mexico City and Hong Kong, the problem has been obvious for a long time. Crowded conditions are undesirable, and adding more people only makes things less desirable. Yet it has never been simply a matter of the human bodies themselves. Even in a poor economy, everybody—every *body*—comes complete with a little bit of space and material: a rice patch here, a thatch hut there, a horse cart or pickup truck, a dirt trail or alleyway or bit of interstate highway. The crowding is not simply of bodies but of *things*, extracted or manufactured things that help constitute GDP. If space were unlimited, there would be no such thing as crowding. Space is obviously limited, however, and more things cause more crowding.

In other words, population and per capita consumption cannot be considered as independent contributors to GDP. Each person contributes to GDP not because he or she is tallied in the national census like some ghost in the cosmos, but because he or she *con-*

sumes, whereby consumption entails the accumulation of goods and the development of locations for the performance of services. Just as clearly, if populations stop growing but things continue accumulating, more space is taken and more crowding occurs, in rich and poor countries alike. Therefore, even economic growth based purely on increasing per capita consumption comes with a price. When the price becomes high enough, as Daly would say, economic growth becomes uneconomic.[21] It costs more than it's worth, causes more problems than it solves, threatens more than it protects. It's a bad deal.

Economic growth may still be a good goal where average per capita consumption is low enough to deprive citizens of a decent home and a decent wage, especially if conditions are not already severely crowded. Surely, however, economic growth has become the wrong goal where per capita consumption is already so high that it comes from increasing levels of unnecessary luxury that displace the good life of family, community and civility. In the United States, Western Europe and Japan, for example, many people have an abundance of material possessions and precious little time to enjoy them.

We have not yet addressed the issue of the truly horrible "bads" that boost GDP. Recalling that anything purchased in the real market (not the speculative stock market or the unreal derivatives market)[22] contributes to GDP, we find that 9/11 "stimulated" millions of dollars of cleanup spending and many billions more on military reprisals and homeland security operations. In other words, the most heinous and devastating act of terrorism in world history created tremendous demand for the production and consumption of goods and services, including some that diversified the economy in very original ways.

Such outlandish "contributions" to the national goal of economic growth led Herman Daly and John Cobb to develop an entirely different approach to accounting for economic welfare. They called it the Index of Sustainable Economic Welfare, or ISEW.[23] The ISEW adds the good transactions and subtracts the bad, and then accounts for non-market indicators of well-being such as education levels and environmental quality. (Following their lead, an organization called Redefining Progress developed the Genuine Progress Indicator, or GPI.) Developing the ISEW was a noble and highly ambitious endeavor, helping to garner a Right Livelihood Award (similar to the Nobel Prize) for Daly, and I hope we live to see it (or the GPI) tallied as carefully as GDP.

However, this is where we should part ways from the many critics of national income accounting who propose modifying GDP *itself* to distinguish the good from the bad transactions. As we have seen, GDP is a very good index of the amount of economic activity within a nation. Whether the transactions are desirable nor not, GDP gives us a good estimate of the economy's *size*. We need to maintain our GDP accounting, in the way it's always been done, for the sake of consistently monitoring economic size.

Consider the analogy of an elephant in a cage. Let's say the cage cannot support an elephant weighing more than 2,000 pounds. An elephant larger than 2,000 pounds will have no room for food, water and excrement, and the cage may even break. If we have a 1,000-pound elephant, we are in good shape regarding the cage. It matters not if the elephant has a bad leg, a bad heart or even bad breath. An elephant in these sickly conditions has a better future than a healthy 2,500-pound elephant forced into the cage. While the small sickly elephant has a chance to recover, the large elephant, though healthy at first, is doomed. The large elephant will either shrink, die a miserably confined death or break the cage, whereupon chaos ensues.

It is certainly a good idea to monitor the elephant's health, no matter how big he is. However, it is just as important to monitor the elephant's *size* no matter how healthy he may otherwise be. If we start with a small elephant and the elephant grows without stopping, eventually the size of the elephant becomes the ultimate issue of concern.

Many economists would see no merit in this analogy because they do not believe the human economy has a cage. We will consider their argument in Chapter 5. For now, letting common sense suffice, let us recognize the Earth as the cage of our global economy. We can monitor economic health with an ISEW or a GPI, but we should continue to monitor size with GDP.

Like the gaining of weight by a small elephant in a large cage, economic growth was surely a good goal when nations were sparsely populated, pollutants were not so toxic and people had few goods and services. But by now it should be just as clear that in a full world economic growth becomes a bad goal. It's also abundantly clear that the world is getting too full, with many nations already in the process of uneconomic growth and with a global footprint too large for the planet to sustain.

At this point, some readers may be left incredulous. After all, in most of the Western world, economic growth has been the most highly esteemed goal of nations since the Great Depression, or certainly since World War II. Roosevelt, Churchill and Stalin vied for it, as did Yoshida in Japan and Mao Zedong in China. Sounds very old, doesn't it? Yet while the world has moved on in so many ways, the growth program has not. If anything, it has intensified. In the United States today, Democrats and Republicans alike support economic growth and compete based upon their abilities to facilitate it. No American president has seriously questioned the goal of economic growth since Dwight D. Eisenhower.[24] Economic growth similarly continues to dominate the political agendas of the European states, Japan, Russia, China, India and most other nations. The only very notable exception is Bhutan in its pursuit of Gross National Happiness. (The King of Thailand advocates the Sufficiency Economy, but much of the Thai polity appears hell-bent on growth.)

Most readers with a dose of common sense will realize by now that economic growth has either become a bad thing at this point in world history or that the crossroads are being approached too quickly for comfort. We need to start applying the brakes. Yet, if you open the pages of a typical economics textbook, you will be told that one of the primary goals, if not *the* primary goal, of

FIGURE 2.3. Federal Reserve headquarters in Washington, DC, (*top*) and the New York Stock Exchange trading floor (*bottom*)—power centers where GDP trends are closely monitored. Credits: (*top*) Dan Smith; (*below*) Ryan Lawler

the economics profession is facilitating economic growth. This wouldn't be a problem if such textbooks were read as often as *The Esoteric Emerson*,[25] but that's not the case. Economics is a common major for many university students, and many other college degrees require the study of basic economics and therefore the reading of basic textbooks. This is especially true for business, by far the most common major in the United States.[26] How many young students question the textbooks they read?

My bachelor's degree was in wildlife ecology at the University of Wisconsin, and it may surprise some readers to hear that I and my fellow wildlife ecology students were required to study economics. In Madison, university curriculum developers recognized that economics affects so much of what happens in the world that even wildlife students, many of whom just want to get away in the mountains, need to know how the rest of the world operates, including the management of budgets and negotiating with commercial interests. We had a choice between microeconomics and macroeconomics, and I chose macro. Since then, however, I have perused many of the leading textbooks in both micro- and macroeconomics. Economic growth, especially growth per capita, is virtually always identified as a primary goal, and usually *the* primary goal in macroeconomics. Some textbooks even identify economic growth, alongside microeconomics and macroeconomics, as one of three major *fields* of economics. That seems like overkill, since economic growth is clearly within the realm of macroeconomics, but it serves to illustrate the emphasis placed on economic growth by the economics profession.

To understand the primacy of economic growth in the economics profession, it will help to briefly consider how the study of economics is organized. Economics is classified as a social science (along with political science and sociology, for example), as opposed to the natural sciences such as physics and biology. The syllables "econ" are derived from the Greek *oikos*, meaning house, or our material environment. Economics deals with the production, distribution and consumption of goods and services. It also deals with the theory and management of economies or economic systems.

The study of microeconomics begins with a look at "economic man," sometimes called *Homo economicus* by conventional economists (and, more often pejoratively, by cynics). Microeconomics is concerned with the behavior of economic man, especially consumers but also producers. For example, what motivates economic man to consume? How much does economic man want to consume, and how does he choose among the available products? The basic premise is that economic man always wants to consume more goods and services, without limit. Of course, he cannot, because his means to consume (such as income) is limited. This constant tension between desire and income is captured in the phrase, "Unlimited wants, limited means." (This opened the door for a creative critic to counter with a book called *Limited Wants, Unlimited Means*, about sustainable tribal economies.)[27]

Microeconomics acknowledges the principle of diminishing returns, as we encountered earlier. It recognizes that, for the average consumer, the desire to purchase a first car, for example, is higher than the desire to purchase a second car, and very much higher than the desire to purchase a fifth car. Regardless of how diminished the desire becomes, however, *Homo economicus* always finds some usefulness, or "utility," in an additional car. Therefore microeconomics introduces the concept of "marginal utility," which is the *extra* utility that an individual receives by consuming one more unit of a particular good.

In more advanced microeconomics, the concept of "disutility" is discussed. The easiest way to understand the concept is to consider goods that are literally consumed, such as eggs. The hungry consumer may truly desire an egg, and a second egg, and maybe even a third or fourth egg. However, the fifth egg, and certainly the fiftieth egg, will cause more problems than it is worth. After consuming four (or forty-nine) eggs, the consumer does not even desire another egg, at least not at the moment. In fact, the consumer desires to *not* eat another egg, because eating another egg will hurt, as Paul Newman memorialized in *Cool Hand Luke*.

However, it is important to realize that the concept of disutility is seldom applied in microeconomics—certainly not in basic microeconomics—and that most of our students study basic, not advanced, microeconomics. Furthermore, there is no acknowledgement in microeconomics that there may be a limit to the number of *types* of goods and services that may enter the marketplace. If economic man has had enough eggs, he may still want some milk. If economic man has had enough food and drink, then there is a virtually endless smorgasbord of non-edible goods to be consumed, or purchased. Beyond that, there are the services, which may include doctoring if too many eggs have been eaten, warehousing if sufficient goods have been purchased and an endless variety of other services somewhat removed from the management of goods, most notably the "softer" entertainment sectors. In other words, economic man does indeed have unlimited wants, according to the microeconomics taught in our schools.

It would not be fair to claim that all economists are true believers in unlimited wants, for no one has taken a comprehensive poll of them. Surely, however, most economists do believe in unlimited wants, and probably some do not. It is more important to realize that economists use conceptual models to describe the behavior of economic man, and that these models simplify reality so that they can be analyzed with simpler mathematical methods.

These models are just algebraic expressions of how economic man behaves. For example, a very simple model of economic man's consumption is $c = y - s$. In this model, c stands for consumption of goods and services, y stands for income and s stands for savings. Economic man receives an income and saves some of it to spend in the future. The rest of it is spent now for the purpose of consuming goods and services. The amount of consumption, then, is calculated as the amount of income minus the amount of saving: $c = y - s$.[28] Simple models like this are assembled to construct theories of economic growth. In fact, usually the entire theory is itself referred to as a model.

It should be emphasized that models are developed in order to simplify things. When simple models are combined to form larger models, however, the math can become exceedingly complex. Therefore, microeconomic models of consumer behavior, especially the models that are widely circulated and understood, seldom incorporate the complicating concept of disutility. In other words, even for the economic modelers who don't believe in unlimited wants, they formally adopt the *assumption* that wants are indeed unlimited, partly because the math would get too complicated otherwise. Eventually, the assumption of unlimited wants is taken so much for granted that it turns into a *belief*, especially among the vast majority of students who are only exposed to the basics.

Microeconomics goes on to look at relationships among supply and demand, wages, prices, profits and the competition that producers face from other producers. It also looks at the market relationships among firms, labor and capital. As for the issue of economic growth, however, the main point is driven home in the first chapter or two of the standard microeconomics textbook: unlimited wants. Therefore, if a student begins a microeconomics course with diligent study habits that decline during the semester (a common occurrence), one of the few lessons the student is likely to take home for life is the concept of unlimited wants, a concept that turns into a belief.

Just as ecology is built upon a foundation of biology, and astronomy upon physics, macroeconomics is built upon a foundation of microeconomics, or at least some microeconomic cornerstones. A typical macroeconomics textbook will begin with an overview or review of microeconomic principles. From there it moves on to the cumulative effects of producers and consumers in the aggregate, usually at the national and international levels. One of the first topics is productivity, which refers to the ability of labor and capital to produce goods and services. From there it proceeds to issues of employment and unemployment, saving and investment, business cycles, inflation, monetary policy, international trade and, of course, economic growth.

Because macroeconomics is built upon a foundation of microeconomics, one of its starting points is "unlimited wants, limited means." Economic man, as perceived in microeconomics, cannot get enough to fully satisfy him. Therefore a national economy consisting of economic men cannot get enough to satisfy *it*. In macroeconomics, economic growth takes center stage and provides the measure of success. Productivity, employment, saving and investment, business cycles, inflation, monetary policy and trade are all analyzed in terms of their effects on economic growth.

The main point of this overview is that our students come out of their economics courses believing economic growth must be an extremely important and unquestioned goal. Micro- and macroeconomic textbooks focus upon pursuing unlimited wants, not ascertaining how limited the means are. When you consider the dominance of economics in the college curriculum, this is no minor point.

One more observation about the college (and often the high school) curriculum is in order. The concept of unlimited wants is taught not only in economics courses. Business has already been mentioned as the most common major in the United States, and usually students get a head start in high school. Business textbooks provide an even more basic overview of economics, especially microeconomics. Business students are taught the concept of unlimited wants early in their curriculum. If they receive a basic overview of macroeconomics, it is focused on the importance and desirability of economic growth. And of course, most business programs require additional studies of economics *per se*, so that business students tend to get a triple whammy of unlimited wants. The concept is introduced in introductory business textbooks, reinforced in microeconomics and extended to the importance of economic growth in macroeconomics. Is it any wonder, then, that the business community considers its role in economic growth as good, unquestioned and even patriotic?

In Part 2, we will consider how the economics profession has described the process of economic growth. With many topics, it helps to consider the historical backdrop in order to understand how current theories have been developed. For the purpose of understanding how economic growth theory has developed, the historical backdrop is absolutely essential. For the sake of entertainment, it also helps that the history of economic growth theory is a fascinating tale of culture, international affairs and political intrigue.

PART 2

THE DISMAL SCIENCE COMES UNHITCHED

CHAPTER 3

Classical Economics: Dealing with the Dismal

> *The superior power of population cannot be checked without producing misery or vice.*
> THOMAS MALTHUS

If you should enter Thomas Jefferson's Monticello home, look back over your left shoulder. There against the wall, next to the doorway, as unmistakable as the Mona Lisa, is the bronze bust of one Francois Quesnay. For me, as a student of early American political economy the sight of Quesnay (pronounced Kay*nay*) at the very entrance to Monticello was like a piece falling into a puzzle. I was thrilled, but not surprised, to suddenly sense the central role that Quesnay had in the thinking of a leading architect of the American Constitution. This helps to legitimize Quesnay as our starting point in discussing the history of economics, and a particularly appropriate starting point for studies of economic *growth*.

Quesnay's upbringing is not entirely clear. He was born in 1694, and biographies portray him as the son of a "ploughman and a merchant"[1] but also as the son of "an advocate and small landed proprietor."[2] He surely grew up observing the countryside and probably working the land to some extent. With diligent self-education, he became a renowned surgeon. By 1749 he found himself serving as a physician in the court of Louis XV of France. To put it mildly, "his life is a model of upward mobility in a supposedly static society."[3]

His interests turned to economics in 1756 when he was asked to contribute a section on farming for an encyclopedia. In 1758 he developed the *Tableau Economique*, or *Economical Table*. The *Tableau* reflects the combination of Quesnay's country-boy agricultural knowledge and his later education in biology and medicine. It resembles a diagram of an economy's "circulatory system," tracing the flow of products, expenditures and income. (As such, it laid the conceptual groundwork for national income accounting.) The *Tableau* demonstrated, at least to Quesnay's satisfaction, that the sole source of economic production was agriculture. Agriculture as the bedrock of the economy would become a prominent feature of the Jeffersonian vision.

Meanwhile the *Tableau* became the foundation for the first identifiable school of economics, "physiocracy," a word which roughly refers to the "rule of nature" or "government of nature." Led by Quesnay and a devoted group of influential followers, the practitioners of physiocracy referred to themselves as *Les Économistes*, the first people to identify themselves as economists.[4] The history books, however, call them the physiocrats.

Much can be inferred about the *Tableau* by considering the language Quesnay used to describe the three classes of people in the economy: the "proprietary class" (landlords), the "productive class" (farmers and agricultural laborers) and the "sterile class" (artisans and merchants). Farmers were the true producers, but in 18th-century France, they seldom owned the land they worked. Artisans, such as clothiers and cobblers, contributed nothing to the economy except goods manufactured from the surplus of farm products, such as wool and hides. Merchants were generally foreigners passing through the economy, buying and selling goods as a service to the farmers, artisans and landlords, neither producing nor manufacturing but making a profit from their transactions.[5] The landlords of this era, often called the "landed aristocracy" or "landed nobility" by historians, were the most fortunate, living easily off rents charged to farmers.

Like the artisans and merchants, the landlords were portrayed as economically "sterile" in the *Tableau*. Their distinctive title, "proprietary class," reflects the fact that landlords were more respected than artisans and merchants. This respect for landlords reflected not only their lineage but also a uniquely French attitude toward agriculture. Perhaps Quesnay also used the term "proprietary class" because he was afraid of insulting the ultimate landlord, King Louis the XV, as "sterile."[6]

Agriculture had long been the pride of France, and still is, along with the associated culinary professions and a refined culture of dining. Although the farmers and laborers did the work, the landlords took an active interest in farming and took pride in their knowledge of the land. They often helped to guide the planting, cultivating and harvesting decisions of the tenant farmers. While

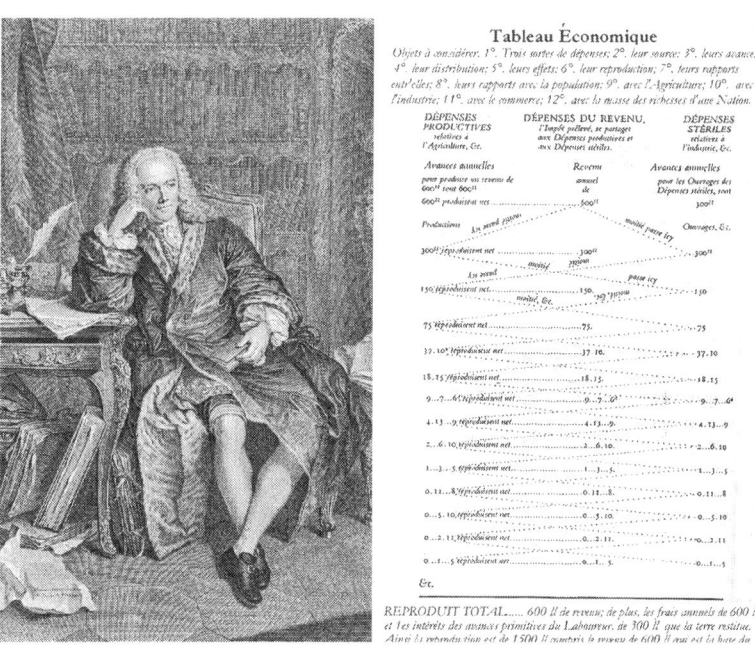

FIGURE 3.1. Francois Quesnay (*left*) and a page from the *Tableau Economique* (*right*). Quesnay could have used the simpler concept of trophic levels (Chapter 7, especially Figure 7.2) to describe the profound importance of agriculture in economic affairs. Credits: Wikimedia Commons; Wikimedia Commons

many were probably arrogant, landlords were the country gentlemen of the 18th century: educated, well-rounded and possessors of rich family histories. Although they earned an easy income, unfairly in the minds of most, they played a proud and important role in French culture.

At the dawn of formal economic studies, economics and politics were considered inextricable. The pioneers of economic study were really engaged in the study of economics *and* politics, or "political economy," but in the history books they are called "classical economists." These classical economists (and predecessors such as Quesnay) combined not only economic and political affairs in their studies; they also tended to have a working knowledge of many of the related sciences. Quesnay had an agricultural background and, through his studies, acquired substantial expertise in a wide range of subjects including biology, medicine, warfare, international trade, political theory and to some degree even wildlife management.[7] As for raw French politics, one could hardly have a better vantage point than Quesnay had in the king's court, where he was called "The Thinker."[8]

Much later, as the "science" of economics developed, economists tried to isolate economic processes from political processes for the purposes of more precise analysis. This development helped to mark the transition to "neoclassical" economics, the mainstream version of economics to this day, and we will elaborate on this transition in the next chapter. For now, however, it should be noted that taking the politics out of economics is like trying to take economics out of politics. After all, the most famous definition of politics is "who gets what, when and how,"[9] while economics is defined in dictionaries as "the social science that deals with the production, distribution and consumption of goods and services." The overlap is almost total. As President Bill Clinton used to say, summing up the biggest political concern, "It's the economy, stupid."

The overlapping of economics and politics was not lost on Quesnay and his followers. They immediately drew two major political applications from the *Tableau Economique*. First, it seemed unfair that the sterile class should benefit so much from the labors of the productive class. The sterile class would not even exist if not for the agricultural surplus of the farmers. Therefore, the physiocrats advocated taxes on land that would redistribute some of the wealth from the landlords to the rest of society, especially the tenant farmers. For example, land taxes could be used to improve wagon roads in the countryside, making it easier for farmers to bring their harvest to market.

If it seems odd that the physiocrats advocated taxing the landlords and not the less-respected artisans and merchants, recall that they saw only agriculture as producing a real surplus. They concluded that only the landlord could really *afford* to be taxed. Taxes on artisans and merchants would simply be converted to higher prices, at the expense of not only the landlords but the hard-pressed farmers. This explains why Quesnay advocated that no taxes other than the land tax be levied. A century later, his "single tax on land" would rise to the forefront of economic politics and policy, with an ironic and astounding impact on economic growth theory, as we will see.

The second political application of physiocracy, following naturally from the first, struck more directly at the establishment of 18th-century Western Europe. In the 17th century, Jean-Baptiste Colbert (1619–1683), a French politician and advisor to Louis XIV, had developed policies for propping up the manufacturing and commercial sectors that benefited the artisans and especially the merchants. Monopoly-granting charters, production controls and protective tariffs came into vogue, instigating a sort of economic warfare among the young European nation-states.

These manipulative government measures came to affect the agricultural sector, too. The most notorious examples were the Corn Laws of England, culminating in the Corn Law of 1815. The Corn Laws restricted imports of grain (wheat, primarily) keeping prices high in an England already suffering a heavy burden of poverty.

In Quesnay's France, the situation was even worse. Agriculture was shackled under a collection of medieval regulations that favored merchants and landlords at the expense of the farmers. The *Tableau Economique* became a tool with which to oppose this interference in economic affairs. The physiocrats began to advocate a policy of *laissez faire*, which literally means "leave alone," at least in today's basic translation from French to English. However, a hundred years later, or halfway between the physiocrats and now, the American Henry George (more about him shortly) thought this about the English use of the phrase laissez faire:

> "Laissez faire!" "Let things alone," has been so emasculated and perverted, but which on their [physiocrats'] lips was "Laissez faire, laissez aller," "Clear the ways and let things alone!" This is said to come from the cry that in medieval tournaments gave the signal for combat. The English motto which I take to come closest to the spirit of the French phrase is, "a fair field and no favor!"[10]

True to his biological expertise, Quesnay saw the products of an economy as blood circulating in a body. As a physician, he knew how interfering with natural circulation was unhealthy for the body. The *Tableau* provided a diagram for a naturally functioning, smoothly flowing economy that was best left alone to serve all its participants in an equitable manner.

To the extent that physiocracy was the first school of economic thought, laissez faire became the first major principle of political economy, and its appeal has endured through the centuries. In the 1980s, for example, laissez faire was a central plank in the platform of the Reagan/Thatcher political economy. The ideological resurgence of laissez faire even helped to bring down Soviet communism.

The curious combination of policies prescribed by the physiocrats—a single tax on land and laissez faire—was more than an effort to keep the economy running naturally, however. It was also a desperate attempt to hold back the tide of unrest among the producers. Tenant farmers and their laborers had become increasingly indistinguishable as heavily-taxed, poverty-ridden peasants.

The single tax would have alleviated that problem and laissez faire would have put an end to the unfair advantage of merchants and artisans in the marketplace. Historians have often wondered if these reforms could have prevented the French Revolution. Perhaps they could have, had they been instituted quickly and strongly. By the reign of Louis XV, however, it was probably too late. As stated by a leading scholar of economic history, "The reforms of Quesnay... were a slight puff of wind countering a developing hurricane."[11]

In any event, laissez faire—Quesnay's major contribution to political economy—was a radical departure from the establishment. As a science of economics, however, physiocracy was far from comprehensive. To borrow from Quesnay's own analogy, the *Tableau Economique* addressed the cardiovascular system, whereas a fully developed economics would have to include the skeletal, muscular and nervous systems.

Through this partly-opened doorway of economic science came Adam Smith (1723–1790), and by the time he left, he had diagnosed the rest of the body economic. As tutor to the young British Duke of Dalkeith, he toured Europe, visited Quesnay and the physiocrats, studied the *Tableau Economique* and returned first to London and then his native Scotland. He began putting his thoughts into writing, and ten years later the result was *An Inquiry into the Nature and Causes of the Wealth of Nations*. *The Wealth of Nations*, published in 1776, was an immediate hit throughout Europe and in the American colonies, where it played a prominent role in shaping the American Constitution.[12] It covered the vast sweep of economic concerns that could be identified at the time, micro and macro. Smith explored the behavior of the individual in the market, describing how the interests of each were worked out such that society as a whole benefited. After all, a man would not purchase something unless he thought it would benefit him, nor would the seller sell unless it benefited him in return. A transaction benefited both, and the accumulation of all transactions benefited society. An "invisible hand," Smith said, ensured that resources were allocated efficiently if the marketplace was free to function naturally. In other words, he agreed with Quesnay that laissez faire should be the guiding economic philosophy.

Smith explored the interactions of the market and government in promoting economic growth, which was clearly a good thing in the relatively empty world of 1776. He described the origins of capital and described how important the accumulation of capital is to economic growth. As we will see in Chapter 5, the accumulation of capital remains a central feature in economic growth theory.

Smith's concept of laissez faire was not a naive one. He was not against any and all government involvement in the economy, but insisted that government's role should be to keep the market functioning as freely as possible. Monopolies, especially, would have to be curbed by the higher powers of government. The system he described, with private ownership of land and capital, wage labor and the invisible hand—assisted when necessary by government—became known as "capitalism." Capitalism, Smith correctly concluded, was the most efficient approach to economic growth and the wealth of nations. In a capitalist system, economic growth became a simple matter of getting evermore land, labor and capital into production.

Smith could not be called a physiocrat, because he denied that agriculture was the sole source of economic production. The popular economic historian Robert Heilbroner called this "one of Smith's greatest insights."[13] Smith argued that labor, wherever it was performed, was the source of economic production. Labor produced goods not only from the land, but in the cobbler's shop and the clothier's factory. The *division of labor*, especially, resulted in a surplus of goods in the manufacturing sectors as well as the agricultural. Meanwhile the capitalist played a key role in generating the division of labor by assembling the factors of production in time and space. Smith's famous pin factory was a tidy example—much tidier than a sprawling farm with its complex seasonal and spatial connections with the natural world.

Frankly, it is difficult to compare the central claim of physiocracy with "one of Smith's greatest insights" without concluding that the issue boils down to semantics. Both arguments are compelling and make sense within their respective frameworks. In the market of Adam Smith, the farmer benefited just as much from the purchase of a hay hook as the manufacturer of the hay hook benefited from the purchase of his daily meal. Yet the physiocrats cannot be so easily dismissed, for is it not just as true that *all* economic activity depended first upon the daily meal? Once agricultural surplus was produced, with or without the hay hook, it became easy to claim that the manufacturer (and even the merchant's services) added value to the economy. But the key phrase is, "once agricultural surplus was produced."

In other words, agriculture is indeed the original *source* of economic production. Yet when there is agricultural surplus, and not everyone has to grow their own food, other activities contribute to economic production, too.

The next classical economist to leave a major mark on the history of economic thought was the Reverend Thomas Robert Malthus (1766–1834). Adam Smith had pointed out that increasing the wealth of nations (economic growth) depended upon a freely operating capitalist system and increasing stocks of land, labor and capital. Smith recognized limits to economic growth when he said that, "…a country which had acquired that full complement of riches which the nature of its soil and climate…allowed it to acquire;…could, therefore, advance no further…"[14] He thought such limits were approximately 200 years in the future: too distant to elaborate upon in *The Wealth of Nations*. Meanwhile, capital accumulation and the division of labor were the keys to economic growth, along with an increasing population to work with the capital and consume the increasing flow of goods and services.

At the time, English politicians were concerned that the British population was not keeping pace with populations on the European mainland. In fact, based upon primitive studies conducted by Dr. Richard Price (1723–1791), a Welsh philosopher, England feared her population was actually declining, threatening her future on the world's economic and military stage. Thomas Malthus turned this fear on its head in *An Essay On the Principle of Population as it Affects the Future Improvement of Society*, published in 1798. He convinced a large following that population growth would always outstrip the agricultural capacity of the land in the long run. Why? Because populations grew exponentially (for example: 2, 4, 8, 16, 32 …) while food production would only grow arithmetically (2, 4, 6, 8, 10 …). As food shortages mounted, malnutrition, disease and eventually wholesale famine would descend upon the human race. The psychological effect of this convincing theory spread far beyond English politics, creating a new mood of wariness and even despair throughout Europe. Economics acquired a new name, "the dismal science."[15]

Today's economists, especially the growth theorists, disregard or even deride Malthus for his prophecies of famine. Ironically, however, his challenging essay became a centerpiece for the concept of scarcity, the primary concern of economists from then on. Furthermore, the verdict on Malthus is not yet in. While his concept of agricultural production was overly simplistic and too pessimistic for 19th-century forecasting, we saw evidence in Chapter 2 for the daunting challenges awaiting agriculture and its consumers in the 21st century and beyond.

Malthus was the first and foremost economist to focus on the limits to economic growth. This was a devastating defection from mainstream political economy, as it crushed the celebratory mood generated by *The Wealth of Nations*.[16] All that had been needed was to let the invisible hand wave its wand and the wealth of nations would grow. Now it looked like growth itself was the biggest threat of all — a real kick in the pants from a suddenly visible foot!

If Malthus had not done enough to sour the mood of Europe, David Ricardo (1772–1823) would handle the rest. Ricardo ignored the historical approach to analysis found in the texts of Smith, Malthus and other classical economists. Instead, he described economic processes in terms of abstract, hypothetical humans who behaved according to certain rules of logic, unswervingly and unnervingly. In other words, without using the term, he introduced the robotic *Homo economicus* — economic man — into classical

FIGURE 3.2. Scenes from the Industrial Revolution, such as "Coalbrookdale by Night" (*above*, depicted by Philip James de Loutherbourg in 1801) accompanied the "dismal science" of Thomas Malthus (*below left*) and other classical economists. John Stuart Mill (*below right*) called for the "stationary state"—not in cultural affairs but in levels of economic activity. Credits: (*above*) The Yorck Project: 10.000 Meisterwerke der Malerei; (*below left*): Wikimedia Commons (public domain); (*below right*) John Watkins (public domain)

economics. He paved the way for a dry, modeling approach that would eventually rule neoclassical economics.

In studies of economic history, much has been made of the disagreements between Malthus and Ricardo. Their worldviews were so dramatically different that disagreement was inevitable. Malthus was an English parson, son of a nature-walking, philosophical eccentric. Ricardo was a stockbroker, son of a shrewd Sephardic Jewish merchant. Malthus was an academic who became the first professor of political economy at a time when professors lived frugal lives. Ricardo went into business at the age of 22, became a wealthy man and retired at the age of 42. Each was an avid writer, but while Malthus toiled in academia until the end, Ricardo became a respected politician in the British House of Commons. They actually became very close friends, but as Heilbroner succinctly put it, "they argued about everything."[17]

Not quite everything, actually. They did not argue about population. They agreed about the tendency of a growing population to push against its food supply, but Ricardo constructed a more elaborate model of the process, with more insight to political economy. In a nutshell it went like this: as population grows, more land must come into agricultural production. Naturally, this land is less productive than the original agricultural land, which was often settled because of its excellent farming potential. Costs of agricultural production increase on these newer marginal farmlands, so food prices rise. The wages of labor—on the farm and in the factory—also rise to meet the increasing costs of subsistence, sometimes shooting above the subsistence level. Wages cannot rise higher than a subsistence level for any significant amount of time, however; they are kept down by the "Iron Law of Wages." Meanwhile, capitalists produce the agricultural implements that allow the farming of the increasingly marginal farmlands, as well as most of the other manufactured goods used throughout society. They innovate and they accumulate capital, becoming more and more productive, but their profits are eaten up by the higher subsistence wages, which keep increasing as the population pushes further into the agricultural hinterlands. Merchants can do little better as their profits depend largely upon the purchasing power of the laborers and capitalists.

Who gains in such a system? The physiocrats would have known immediately! The landlord gains in such a system, because as food prices continue to rise, rents on land continue to climb. Indeed, the well-established landlords with the most productive agricultural lands become extravagantly wealthy. In Ricardo's view, the never-resting capitalist has suffered the greatest inequity.

This is the dismal view Ricardo is remembered for, because he emphasized the Iron Law of Wages, the unrelenting pressure on the capitalist and the inevitable, unearned wealth of the landlord. As John Kenneth Galbraith pointed out, however, those who dig deeper into the writings of Ricardo will find that he also believed in the possibility of technological progress and capital accumulation to keep conditions from becoming truly Malthusian. In other words, we see in Ricardo some of the first hints of neoclassical growth theory (Chapter 5) and the notion of perpetual economic growth.

One more of Ricardo's contributions is highly relevant to the development of economic growth theory. It is called the principle of "comparative advantage." This is the principle that has done more than anything else in economics to support the practice of international trade, and it is just as relevant today as it was when Ricardo crafted it.

To understand the principle of comparative advantage, we first consider the simpler principle of "absolute advantage," which Adam Smith had written about. Absolute advantage occurs when one nation is able to produce a good more cheaply than another nation. For example, if the average farm in Ukraine produces wheat at a cost of $1.50 per bushel, and the average farm in the US produces wheat at $1.00 per bushel, then the US has an absolute advantage over Ukraine in wheat production. Now consider a second commodity, such as butter. If Ukraine butter costs $3.00 per pound and American butter $1.50 per pound, the US also has an absolute advantage over Ukraine in butter production. (While these are hypothetical examples, the US can generally produce agricultural commodities more cheaply than other nations due to its natural resources, technological advantages and economies of scale.)

At first glance, it may appear as though the US would have little reason to trade with Ukraine. The US could trade butter for Ukrainian wheat, but it can already produce wheat more cheaply than Ukraine. Similarly, the US could trade wheat for Ukrainian butter, but it can already produce butter more cheaply than Ukraine. Ricardo wasn't fooled, however. He saw that it was mutually beneficial for both countries to specialize and trade.

In Ukraine, one pound of butter costs the same amount to produce (approximately $3.00) as two bushels of wheat. In other words, the production of one pound of butter means foregoing the production of two bushels of wheat in Ukraine. In a sense, it "costs" Ukraine two bushels of wheat to produce one pound of butter. In the US, meanwhile, one pound of butter costs the same amount to produce (approximately $1.50) as 1.5 bushels of wheat, so it costs the US only 1.5 bushels of wheat to produce a pound of butter. Therefore, the US is *relatively* better at producing butter than Ukraine, and is said to have a *comparative* advantage in the production of butter.

Meanwhile, in Ukraine, one bushel of wheat costs the same amount to produce ($1.50) as half a pound of butter, so that it costs Ukraine half a pound of butter to produce one bushel of wheat. In the US, one bushel of wheat costs the same amount to produce ($1.00) as two-thirds of a pound of butter, so that it costs the US two-thirds of a pound of butter to produce one bushel of wheat. Therefore, Ukraine is relatively better at producing wheat than the US, and Ukraine is said to have a comparative advantage in the production of wheat.

Because these relative or comparative costs differ between butter and wheat, it is mutually advantageous for both countries to specialize and trade, even though the US has an absolute advantage in both commodities. If the US specializes in butter and Ukraine specializes in wheat, the two nations together will produce more butter and wheat in the aggregate. In other words, the two nations combined will be wealthier.

For example, if Ukraine and the US each started with a budget of $45.00, and refused to specialize or trade, Ukraine would be able to produce ten bushels of wheat and ten pounds of butter. The US, with its absolute advantage in both commodities, could produce 18 bushels of wheat and 18 pounds of butter. The total, international production would be 28 bushels of wheat and 28 pounds of butter.

However, if each nation exercised its comparative advantage and specialized, then Ukraine would produce 30 bushels of wheat ($45.00 divided by 1.50 per bushel) and the US would produce 30 pounds of butter ($45.00 divided by 1.50 per pound). International production would therefore increase to 30 bushels of wheat and 30 pounds of butter. If each nation wanted a balance of wheat and butter, they would trade (US butter for Ukrainian wheat and vice versa). Assuming the terms of trade were fair, each nation would gain somewhat from the surplus, although the US would be able to demand more of the surplus because of its absolute advantage in both commodities. In other words, each nation would be better off than if it had produced all its own commodities and refused to specialize or trade.

Ricardo's insight had an immediate impact on the politics of international trade. The farms of England were, at best, only slightly more productive than the farms of many European nations, and for the most part were less productive than those of Spain and especially France. However, the English were more advanced industrially than any other nation. As the principle of comparative advantage spread through the English political economy, the landlords lost much of their clout to the capitalists and merchants. Parliament loosened its protection of English agriculture and the nation began to trade its manufactured goods for agricultural produce. The global economy produced more, and all nations who managed their trade for comparative advantage were better off than they had been.

A full world provides a different perspective on the merits of free trade, of course. Seeking and practicing more comparative advantage in a full world is like milking the planetary cash-cow more completely. If the cow is going dry, it's time to stop milking! But the world was far from full in Ricardo's time.

The world as a whole was far from full, but by the early 19th century when Malthus and Ricardo wrote, industrial capitalism had been running roughshod in the cities of England, especially London, for nearly half a century. Conditions for workers were appalling: crowded, hectic, dangerous, noisy and noxious. Workers were pushed to the breaking point, often working 18-hour days, 6 or 7 days a week. Some of their more privileged countrymen observed the workers' plight and felt obligated to improve their lot in life. In some cases these conscientious souls also developed a resentment toward the capitalists and landlords who took advantage of the powerless workers. They even began to question why some in society should ever obtain the privilege of owning large stocks of capital and thusly attaining an immense economic and political advantage over others. In other words, the concept of communism began to take root, although it was not well-defined. Those who touted a more socially- or community-minded economic system at this point in history are sometimes classified in the history books as "utopians." In *The Worldly Philosophers* Heilbroner called them "utopian socialists." They had little to say about economic growth; their focus was the injustice brought upon the workers by the capitalist system. It is important to acknowledge their presence, however, because they set the stage for two of the most brilliant economists of all, John Stuart Mill and Karl Marx.

John Stuart Mill (1806–1873) stands out as the most knowledgeable and scholarly of the classical economists. This was partly a matter of heritage because his father was James Mill, a historian and philosopher who was a close friend of Ricardo. The elder Mill was not only a renowned scholar himself, he drove his son to excessive levels of study. John Stuart Mill is said to have begun learning Greek when he was three years old, and by the time he was thirteen had written a survey of the entire field of political economy. The pace and discipline eventually took its toll psychologically, but after somewhat of a crisis he pulled through and flourished in scholarship for the rest of his life. For our purposes, his major contributions were found in *The Principles of Political Economy* (1848), the most comprehensive tract on economics up to that time, and perhaps to this day.

One such contribution was Mill's critique of the Malthusian and Ricardian assumption that, when it comes to economic behavior, people simply act like robots. Ricardo's model, especially, had assembled a cast of pawns who were seemingly bound by fate to an unjust, miserable outcome, with a few lucky landlords winning the game hands down. Mill gave the reader hope by emphasizing the difference between production and distribution.

Mill said that the production of goods was indeed ruled by iron-clad laws. Combining a certain amount of labor with a certain unit of capital on a certain parcel of land could produce only one thing. A forest could produce only certain types of lumber. A lamppost factory was tooled and calibrated in such a way that limited it to the production of a certain type and number of lampposts. There was little mystery in the process of production, although change was frequent as new inventions and methods were discovered. The laws stayed the same; technological progress simply allowed for the laws to be applied in new scenarios.

The distribution of wealth, on the other hand, was determined wholly by human preference. Culture, religion, laws, education and social norms all played a role in how humans distributed the products of farm and factory. Economic man was not a robot after all; how could anyone have thought so? Men and women were conscious and cultured, sometimes caring and sometimes crass. If society didn't want to play by the unfair rules of Ricardo's game, it didn't have to. People could share the land or at least the pie if they wanted to. The plausibility of this was good news to the throngs who were mortified by Malthus, Ricardo and their fatalistic disciples.

Pertaining to economic growth, the major contribution of Mill was his concept of the "stationary state."[18] The stationary state would be a natural outcome in a society that decided to take control of its economic affairs. The key was a stable population. Mill wasn't talking about government intervention; he thought an educated public would voluntarily lower its birth rate such that population would stabilize. The conclusions of Malthus and Ricardo had been based upon ever-increasing populations. Recall that in Ricardo's more sophisticated version, it was increasing population that allowed and motivated the capitalist to hire more labor and accumulate more capital. The Iron Law of Wages kept the wages of labor bouncing on a dirt floor of mere subsistence, yet the capitalist couldn't get ahead either, because the subsistence level of wages kept increasing as marginal farmlands (where it cost more to produce) were brought into production. With a stabilized population, the labor force would stabilize and so would the subsistence level of wages. The capitalist might still want to increase his capital stock, but with a stable population an increasing capital stock would result in a rising wage. In this case wages would rise higher than the subsistence level; the Iron Law of Wages would be broken. In the Ricardian model of increasing population, the capitalist was trapped between the subsistence wage and the landlord's ability to charge ever-higher rents. Therefore, he was continually forced to accumulate more capital just to keep afloat. With Mill's stable population there was no such pressure to accumulate and, furthermore, accumulation would go unrewarded. The stationary state would be the result.

This is the version of Mill's stationary state that one may readily find in a typical reader on economic history. By digging a bit further into the *The Principles of Political Economy*, however, one finds that Mill was onto something much more. In fact, he was envisioning a *full world* à la Herman Daly in which the stationary state was not only likely but desirable. Here is some indisputable evidence:

> It must always have been seen, more or less distinctly, by political economists, that the increase in wealth is not boundless: that at the end of what they term the progressive state

[economic growth] lies the stationary state... It is not good for a man to be kept perforce at all times in the presence of his species.... Nor is there much satisfaction in contemplating the world with nothing left to the spontaneous activity of nature; with every rood of land brought into cultivation, which is capable of growing food for human beings; every flowery waste or natural pasture plowed up, all quadrupeds or birds which are not domesticated for man's use exterminated as his rivals for food, every hedgerow or superfluous tree rooted out, and scarcely a place left where a wild shrub or flower could grow without being eradicated as a weed in the name of improved agriculture. If the earth must lose that great portion of its pleasantness which it owes to things that the unlimited increase of wealth and population would extirpate from it, for the mere purpose of enabling it to support a larger, but not a happier or a better population, I sincerely hope, for the sake of posterity, that they will be content to be stationary, long before necessity compels them to it.[19]

As a certified wildlife biologist who came into economic studies from the perspective of wildlife conservation, I find this portion of *The Principles* fascinating. In talks to wildlife professionals at conferences and universities, I like to quiz the audience by quoting from this portion and asking, "Who said this?" Invariably a significant percentage of the audience answers, "Aldo Leopold." Aldo Leopold is considered the father of wildlife ecology and management, and no field of study is more identified with an individual than wildlife ecology is with Leopold. Leopold's *Sand County Almanac*, with its philosophy of the land ethic, is to wildlifers what *The Principles* was to political economists, and wildlifers quote Leopold like politicians quote Lincoln. When wildlife biologists aren't sure what to do with an issue, the question is invariably asked, "What would Leopold say?" It is remarkable that an all-time master of political economy spoke in terms readily identifiable with the icon of wildlife conservation.

But times have changed. Walt Whitman Rostow (1916–2003), an economist who served as a special assistant for national security affairs for presidents John F. Kennedy and Lyndon B. Johnson, took the position that "Mill was the first major environmentalist," yet concluded: "it should be underlined that his stationary state implies only a fixed population; technological change could proceed elevating real income per capita."[20] Rostow was wrong. Mill clearly saw all economic production as rooted ultimately in the land. The last sentence in the lengthy Mill quote above reveals that Mill saw "the unlimited increase of wealth and population" as forces that "extirpate" the Earth's "pleasantness." His "sincere hope" was that people would bring their *wealth and population* (not just population as Rostow stated) under control "before necessity compels them to it." For Mill, there was indeed a limit to economic growth, whether based on population growth or growth in per capita consumption.

Mill connected the stationary state to the free will of humans. He saw humans as cultural, political and spiritual, not just economic, beings. He envisioned society turning from a mindless procession of economic growth to a mindful ordering of its cultural and political affairs, with liberty and justice for all, as well as tending to its spiritual needs. Mill refined his vision for this higher plane of society in a later book, *On Liberty*. Like Adam Smith, he was as much a moral and political philosopher as an economist: a truly classical, well-rounded, political economist.

In the history of economic thought, Mill is considered the great synthesizer of classical economics. Therefore, his comparison between capitalism and communism, or at least the prospects for communism, is worth noting: "If the choice were to be made between Communism with all its chances, and the present state of society with all its sufferings and injustices...if this or Communism were the alternatives, all the difficulties, great or small, of Communism would be as dust in the balance."[21]

On the other hand, this seemingly strong language was so diluted by disclaimers and counteracting speculation that it is difficult to know where Mill really stood. Furthermore, "Communism" was still the unshaped, innocent brainchild of the utopian socialists. Mill's *The Principles of Political Economy* was no communist manifesto as we might imagine one with the hindsight of history. Ironically, the real thing, *The Communist Manifesto*, was coming off the printing press at the very same time (1848), which brings us to Karl Marx (1818–1883).

Without a doubt, Marx was the most famous, infamous and influential radical of the 19th century, in or out of economics. He was a rebellious and mostly unproductive student, but eventually immersed himself in philosophy, political economy and history, synthesizing these disciplines from all angles. He was unquestionably a genius, probably the best-read economist of his time (with the possible exception of Mill), and a perfectionist as an author. In developing his thoughts, he also had an enormous advantage over the likes of Adam Smith. Not only did he have *The Wealth of Nations* and the other classics under his belt, but a century's worth of capitalist history to draw upon. He didn't like what he saw, but for different reasons than Malthus and Ricardo.

For one thing, the landed aristocracy was rapidly becoming a shadow of its 18th-century royalty. True, there was a new breed of land baron coming to power in the United States, but Marx's perspective was European. In any event, in Europe *and* the United States it was the age of the industrial capitalist, landowning or not. Marx described at length how capitalism had the unrelenting tendency to concentrate wealth in the hands of fewer and fewer capitalists, the "expropriators." He concluded that capitalism was doomed because the working class, or proletariat, would suffer only so much abuse and then revolt. They would take from the capitalists by force, and they should! This was not something the mild-mannered Mill would have ordained. Nor would Mill have thought such a process inevitable.

Marx thought Mill was naive on the prospects for mankind to distribute wealth fairly, but he didn't think economic man was a Ricardian robot, either. His theory was more complex, critical and cynical. Marx is considered the great-grandfather of a tradition in political science called critical theory.[22] He thought governments were ruled by capitalist interests that would suppress the efforts of the proletariat to act in its own interests—for as long as possible, at least. But the suppression could not last. For Marx, a communist revolution was a juggernaut just around the corner in Europe. There was so much social unrest over economic conditions; we will never know how close the juggernaut came to rounding that corner in Europe. We do know the juggernaut veered eastward instead.

Marx did not elaborate on how the proletariat would reorganize and redistribute capital, other than to claim it would be communally owned and operated. His most fantastic claim was that eventually there would be no need for the state. In the interim, however, a new form of government, a socialist government, would have to keep things in order.

The Communist Manifesto was a relatively simple *political* call to the proletariat that contained some of the most brazen, anti-establishment rhetoric ever published. He summarized his vision of a communist revolution with the immortal words (notorious to some, heroic to others), "The expropriators are expropriated!"

Marx's highly complex economic *theory*, on the other hand, was published from 1867 to 1910 as the four volumes of *Das Kapital*. The combination of the two works—one a short political tract, the other an exhaustive theory of political economy—was so powerful that a mighty struggle ensued for the minds of statesmen worldwide. A major communist revolution did not occur in Europe as Marx predicted, but eventually his ideas took root to the east. The Russian revolution of the early 20th century was communist to the core and exceedingly bloody. For our purposes the most relevant outcome was that, at one point in history, two of the three global superpowers—the Soviet Union and China—organized their economic systems in accordance with Marxists precepts. This polarization of power precipitated a political, economic and military race: the Cold War.

As for the eventual, "inevitable," stateless society? The Soviet Union struggled with the details of communist government until it collapsed, leaving Russia and its satellites to start over under various shades of capitalism. China is still ruled by the Communist Party, but it's hard to know what that means when the Chinese economy is more capitalist by the day. Meanwhile the Cuban, North Korean and a few other communist governments continue their struggles to compete in an evermore capitalistic world.

Yet our primary concern is not with the supposedly inevitable transition from capitalism to communism, but rather the transition of economic growth from a good thing to a bad thing. Did Marx have any vision of long-term economic growth? It is difficult to tell. He had seen rapid technological progress in agricultural production and spurned Malthus for his seemingly simplistic view of population growth. And one of the more curious aspects of *Das Kapital* is that, with all the venom Marx spit at the capitalist, he celebrated the Industrial Revolution for the productive powers it unleashed. It was this industrial power that, if only wrested from the hands of monopolizing capitalists, would produce the economic surplus required to (somehow) make government unnecessary. Marx predicted a breakdown of the capitalist system, but the breakdown would be caused by the excessive *concentration* of capitalist wealth rather than an absolute *shortage* of wealth. Because he did not provide a clear picture of what would follow the breakdown, however, it is difficult to determine his long-term view on economic growth. The question is whether Marx saw a communist economy as stationary or perpetually growing. Today, many students of Marx argue on both sides, and Paul Ormerod put it most safely, "Marx did have doubts about the ultimate sustainability of growth."[23]

As for the fate of capitalist nations, Marx failed to acknowledge that governments might come to grips with the destructive forces of capitalism. The United States, for example, is founded upon a constitution that allows for not only a capitalist economic system, but establishes a democratic political system. The United States has long been classified as a capitalist *democracy*, and it did respond to many of the injustices of 19th-century industrial capitalism, to some degree at least, as did European governments. Theoretically, a democratic government could also steer its capitalist economy safely through the transition from an empty world to a full world.

Theoretically, in other words, a democratic government could bring about a steady state economy even with many of the trappings of capitalism intact. In any case, the communist nations, true to Marx's vision or not, were just as preoccupied with economic growth as their capitalist rivals. The lone exception may be Cuba, which seemingly has accepted a non-growing economy (in the face of economic sanctions), instilling citizen pride with performance measures such as education levels and health care regardless of GDP growth.

As with Malthus, the verdict is not yet in on Marx. Just as Malthus jumped the gun in predicting widespread food shortages in a relatively empty Europe without realizing the tremendous gains to be made in agricultural productivity, Marx predicted a European communist revolution without realizing the gains to be made in democratic governance. We must admit, however, that numerous Malthusian scenarios have played out on the planet already, at least at the national or regional level. So have communist revolutions. Furthermore, as we saw in Chapter 1, a widespread, potentially global Malthusian outcome is far from ruled out. Marx thought Malthus was naive, but to the extent that economic growth in a capitalist system dominates politics, building toward a Malthusian scenario, one could say that Marx proves Malthus correct. As they say, the ironies never cease.

For the purposes of avoiding a Malthusian outcome, it was perhaps an unfortunate coincidence that another major defector in economic thought was eventually eclipsed by Marxist movements and countermovements. The defector's name was Henry George (1839–1897). Just as Americans know little of Marx, Europeans know little of George. In fact, even *Americans* know little of George. Our great-grandparents knew plenty about him, though. Our lack of knowledge has less to do with the true place of Henry George in history than the efforts to snuff him out of our memories, as we are about to see.

CHAPTER 4

"Neoclassical" Economics: Dealing with the Devil

> *Neoclassical economics is the idiom of most economic discourse today. It is the paradigm that bends the twigs of young minds. Then it confines the fluorescence of older ones, like chicken-wire shaping a topiary.*
> MASON GAFFNEY

THE CLASSICAL ERA of economics came to a close during the latter decades of the 19th century. It seemed like all the great thoughts had been thought, all the core principles discovered, all the big issues debated. The Industrial Revolution had issued its most violent contortions, the West had settled into the age of capitalism, and by early in the 20th century, the East had expropriated its expropriators. Now it was time to wait and see what the outcome might be.

Much of classical economics had amounted to addressing the principles laid out in Adam Smith's *The Wealth of Nations*. Malthus, Ricardo, Mill and a few others contributed original and important perspectives, differing on key points, building upon societal developments. Marx, on the other hand, had challenged the entire economics establishment, in and out of academia. His followers were not to be called classical economists but Marxists. Henry George was about to present another serious challenge and a unique solution; his followers would be called Georgists. Here and there were smaller factions of economic thought—American institutionalists, anarchists, Fabian socialists—but the majority of economists in the Western world kept to the general path laid out by the classical economists and entered the age of "neoclassical" economics.

The transition from classical to neoclassical economics was not sudden, not even by the standards of slow and cautious academic movements. The first use of the phrase "neoclassical economics" may be obscured in the annals of economic literature, but Thorstein Veblen is generally credited with coining the phrase, aptly enough in 1900.[1] Those we view today as the pioneers of neoclassical economics still thought of their studies as "political economy."

A typical dictionary tells us the root "neo" has three distinct meanings. First and most simply, it means new, as when "neophyte" is used to identify a convert or beginner. As applied to anything classical, however, "new" would be oxymoronic, for how can something be simultaneously new and classical? Secondly, however, neo may refer to something new and different, but rooted in the original. This usage works with neoclassical economics, which is rooted in classical principles yet is a newer and different version of economics. Thirdly, neo may be used to identify or connote the "New World" (Western hemisphere), as in "neotropical." As applied to economics, then, neoclassical would refer to the modification of classical economics to fit with the needs, concerns and events of the New World. This is not how economists consciously use the phrase, but it is an interesting coincidence that the influence of New World, American economists began in earnest with the transition from classical to neoclassical economics.

A quick perusal of the literature will reveal numerous demarcation points for the neoclassical transition. The simplest ones are based on the tenures of particularly influential economists, such as Carl Menger at the University of Vienna, John Bates Clark at Columbia University, Alfred Marshall at Cambridge or, at the latest, Paul Samuelson at the Massachusetts Institute of Technology. These designations run, therefore, from the 1870s to the 1940s.

A more sophisticated analysis breaks the neoclassical movement into an Anglo-Saxon tradition, which lumps together the American and English schools of thought, and a Continental tradition, which lumps together several schools of thought from the European mainland. As a portent of the major theme of this chapter, we should note that at least some authorities have classified the American school as "American apologists."[2] Their apology was for the capitalist system, heavy concentrations of wealth and the transition to their brand of neoclassical economics.

The Anglo-Saxon and Continental traditions are both rooted, more or less, in the Marginalist Revolution of 1871–1874. The details of the Marginalist Revolution are unnecessary for our purposes, but there are some key points to be aware of regarding its impact on economic growth theory and policy. The Marginalist Revolution gets its name from the concept of marginality or marginal units. As noted in Chapter 2, a marginal unit is simply an *extra* unit of something, such as cost, utility or productivity. For example, a consumer always finds some usefulness, or utility, in a good or service. The marginalists introduced the concept of *marginal* utility, which is the *extra* utility that an individual receives by consuming one more unit of a particular good. This, they showed, was a crucial factor in determining the price of the good.

In classical economics, a great deal of attention had been paid to prices: how they were derived and how they affected the standard of living. The classical economists hadn't figured out the mysteries of marginal utility, however. They acknowledged that utility could affect market prices in the short term, but they believed that the long-run, "natural" price of a good could only reflect the labor costs that went into producing it. This viewpoint was referred to as the labor theory of value, and it led to some extremely complex approaches to accounting for labor costs. For example, if a farmer purchased a newly developed seeder, was the price supposed to reflect only the labor of those who produced the seeder, or also the labor of those who had invented the seeder and all its constituent parts? What about the labor of those who advertised and distributed the seeder? Needless to say, this approach to ascertaining appropriate prices produced some highly inconsistent, confusing and controversial results.

The concept of marginal utility was a vast improvement because it clarified how *demand* influences prices. Increasing demand causes prices to increase. Demand, meanwhile, reflects the utility of the good being demanded. As the marginalists realized, however, demand changed with the amount consumed. In particular, demand is subject to diminishing marginal utility. As more of a good is consumed, less is demanded. This was the missing link for a realistic understanding of prices in the market.

From 1871 to 1874, principles of marginality were applied with increasing sophistication to the workings of the market. Carl Menger, Léon Walras and William Stanley Jevons were the primary architects of this new model of the market. Walras and Jevons were eminent mathematicians and set the lasting precedent—whether good or bad—for the application of complex mathematics to economics. It didn't take long for Walras to develop a model of "general equilibrium," a situation in which all markets in the economy (wheat, shoes, houses, etc.) have "cleared," meaning the demand for all goods has equaled the supply. In general equilibrium, therefore, Say's law is upheld.

One of the key principles of general equilibrium is that, as the marginal utility of a good diminishes with consumption, consumers turn to purchasing other goods that offer higher marginal utility. In this way, not only do all markets clear but the allocation of resources used to supply the goods maximizes utility at large. General equilibrium is viewed as the most efficient market scenario. Walras also demonstrated how, in general equilibrium, prices throughout the entire economy at any given time could be determined using a long series of mathematical equations.

Certain assumptions had to be made for this general equilibrium to hold, most notably perfect competition among producers. This and other rigid assumptions have been the focus of critics who have found general equilibrium an artificial and misleading model of market reality. Furthermore, general equilibrium was a "static theory," meaning it considered the market at a single point in time, offering little insight to trends and projections. The excessive mathematicization and abstraction from reality became trends that eventually led the Nobel Prize-winning economist Wassily Leontief to lament, "Departments of economics are graduating a generation of idiot savants, brilliant at esoteric mathematics yet innocent of actual economic life."[3] Nevertheless, general equilibrium remains a holy grail of microeconomics to this day.

The last sentence reveals another important point about the Marginalist Revolution. We saw that the classical economists were really scholars of *political* economy, concerned with the vast breadth and depth of social and political factors that influenced whole economies. To the extent they focused on economics as such, their approach was macroeconomic: the wealth of nations, population growth, international trade. The marginalists, in contrast, built their models from the bottom up, considering the principles of supply and demand "at the margin"—by individual people for individual goods. They rapidly constructed models with more than one, and then many, goods and services. Eventually Walras took in the entire economy, but it was a highly abstract exercise in which the precise identities of the goods in question were reduced to algebraic symbols. In some ways, the approach of the neoclassical economists was reminiscent of Ricardo's robotic scheme, except now even the moving robots were gone, replaced by inanimate symbols. There were no social classes to be abused, enriched, mobilized or opposed, just producers allocating resources and consumers maximizing utility. The essence of general equilibrium theory can be found in Walras's *Elements of Pure Economics* (1874):

> In fact, the whole world may be looked upon as a vast general market made up of diverse special markets where social wealth is bought and sold. Our task then is to discover the laws to which these purchases and sales tend to conform automatically. To this end, we shall suppose that the market is

perfectly competitive, just as in pure mechanics we suppose to start with, that machines are perfectly frictionless.[4]

The Marginalist Revolution spawned the transition not only from classical to neoclassical economics, but also from a focus on macroeconomics to microeconomics. In other words, neoclassical economics is largely about microeconomics and, as such, has little to say about the process of economic growth. Instead, neoclassical economics, rooted in microeconomics, focuses on the efficient allocation of resources and the maximizing of utility. The neoclassical version of macroeconomics, if there is such a thing, is an extension of microeconomics supplemented by principles developed by John Maynard Keynes, including more of a focus on the monetary sector (as opposed to the "real" sector), and gradually augmented with considerations of technological progress (to be discussed later in this chapter). Some scholars consider macroeconomics to be entirely outside the realm of neoclassical economics.

But for neoclassical economics, the Marginalist Revolution was just the beginning. Political economy and classical economics were in no way dead. It would take other developments in economic thought to warrant a name change from classical to neoclassical economics.

Onto the stage strode Henry George, an American who unwittingly assisted the transition in a most ironic way.

George's *Progress and Poverty* was published in 1879. It had a huge impact in North America and Australia, and for a while it had many followers in Europe too, especially Great Britain and Ireland. In fact, George once had far more worldwide support than Marx, who tended to polarize even the critics of capitalism. Philosophers the likes of Leo Tolstoy, prime ministers including David Lloyd George, and revolutionaries like Alexander Kerensky were among the champions of "Georgist" political economy.

We might think of George like this: what Marx was to labor, George was to land and what Marx was to the capitalist, George was to the land baron. Marx saw the capitalist system aligned against the laborer, with capitalists grabbing an increasingly unjust and dangerous share of society's wealth. George also saw the system as oppressing the laborer, but to George, it was not because of the concentration of capital but rather the concentration of land in the hands of few. To Marx, the capitalist's extraction of wealth from the toils of the proletariat was the greatest injustice of capitalism. To George, the landlord's unearned wealth from rents (that rose incessantly as populations and businesses expanded) was the biggest travesty. Marx called for a communist revolution, George called for…a tax on land!

In a nutshell, George's argument was this: wealth consists of tangible goods, and an increase in these goods represents an increase in wealth. The increasing goods are readily distinguished from land, because land cannot increase in quantity. As populations grow, land rents increase, but since the land itself does not grow, the "common wealth" does not increase. Instead, increasing land rent simply amounts to an ever-widening maldistribution of wealth, which moves from the tenant (who is inevitably a laborer) to the landlord. More money may be spent on land, but it is money earned by the toil of the laborer, then delivered into the lazy hands of the landlord.

To better understand George's argument, it helps to understand how George and the classical economists used the term "rent." Rent, as we generally think of it, refers to the income landlords receive for the use of buildings or machinery that occupy an area of land. In the Georgist sense, however, rent refers to that portion of this income that results solely from owning the land, and is, in some fundamental sense, "unearned." For example, a landlord may lease a property with a house trailer for $1,000 per month. A neighboring landlord may lease an otherwise identical property with no house trailer for $600; a tenant wishing to live there may also rent a house trailer for an additional $400 and move it to the land in question, costing him a total of $1,000 per month. All else equal, this scenario indicates that the first landlord has invested in a capital good (a house trailer) that is worth about $400 per month. The unearned rent, according to George, is $600. In other words, rent is that portion of income that results simply from the act of owning the land. If the rent was taxed at a rate of 100 percent, the landlord would pay a tax of $600 per month. Any rental income would have to be earned by the landlord. He would have to improve the property in some way, for example by tilling a garden plot, improving a driveway or installing a house trailer. There would be no free lunch.

George was not the first to note the social injustice of unearned income for landowners and the tendency of rents to increase faster than tenants' incomes. We began Chapter 3 with a look at the physiocrats, who identified the rent-taking of the "proprietary class" as a fly in the ointment of the French economy. Throughout the 19th century various classical economists came back to this theme, most notably Ricardo. George was not even the first to propose a tax on land. Again the physiocrats set that precedent, and clearly they felt strongly about it, because otherwise they championed laissez faire.

Two things distinguished George, however. First, he wrote with a passionate style. His rhetoric was vaguely reminiscent of *The Communist Manifesto* but, unlike the atheistic Marx, George infused *Progress and Poverty* with Christian exhortation. This was a potent mix, given the Protestant ethic that permeated the American agricultural economy. Consider this salvo from Chapter 26:

> Can it be that the gifts of the Creator may be thus misappropriated with impunity? Is it a light thing that labour should be robbed of its earnings while greed rolls in wealth—that the many should want while the few are surfeited? Turn to history, and on every page may be read the lesson that such wrong never goes unpunished; that the nemesis that follows injustice never falters nor sleeps. Look around today. Can this state of things continue?…Nay; the pillars of the state are trembling even now, and the very foundations of society begin to quiver with pent-up forces that glow underneath.

The struggle that must either revivify, or convulse in ruin, is near at hand, if it be not already begun.

The second thing that distinguished George was when and where he wrote. If ever the time and place had come for a serious land tax, it was the late-19th-century American West. A handful of wealthy Americans—railroad, timber and cattle barons—had managed to amass millions of acres, often by luck, trickery and brute force. Other types of landlords dominated different political regions of the United States. In the East, especially, these landlords tended to be the very same capitalists who invested their profits in land as well as capital. After all, land rents were sure to rise amidst the floodtide of European immigrants, while industrial profits were always at risk of the invisible hand. Capitalism could be a stressful and highly competitive occupation; land lording required little more than buying the lands, waiting a bit and collecting the rents.

Many immigrants had fled Europe because of oppressive landowning regimes in their native countries. Aristocracy, vestiges of feudalism and Roman Catholic patronage had kept masses of Europeans in a state of landless peasantry. When ships set sail for the New World, the immigrants were ready for a new life. They did not want to settle for a new form of peasantry, and there were plenty of descendents of earlier immigrants already populating the United States with similar sentiments. It is not so surprising, then, that when *Progress and Poverty* was published, it ignited a powder keg of pent-up frustration. By some accounts, it sold more copies than any book published through the first decade of the 20th century with the exception of the Bible.[5] Many of George's ideas were rolled up into the 1890s Populist Movement and then the Progressive Movement, which dominated Republican and Democratic Party agendas from 1902–1919.

Ideally, according to George, *only* land would be taxed. The factors of production were land, labor and capital. It was not right to tax the wages of labor because the laborers earned their keep, and taxing them would only discourage workers from performing diligently. It was foolhardy to tax capital (or capital gains), because capital investment helped make the economy more productive. Landlords, however, were collecting unearned money. Taxing them would serve justice. A substantial fringe benefit would be to discourage the land speculation that often caused heartbreaking boom-and-bust cycles in the American West.

Taxing the landlord and the speculator makes a lot of sense, and George's proposal bore some fruit. Property taxes became a major source of revenue for local governments in the United States, and the federal income tax targeted land rents early in the 20th century. Yet local property taxes have given way to sales taxes, and the federal income tax has honed in on wages more than rents. George's dream of a tax on land, and land only, never came close to fruition. Furthermore, George wanted *all* rent from land taxed, with the revenue spread across society like a blanket of security, not just enough to keep the government running at a minimal level.

Why spend so much time talking about land taxes in a book about economic growth? The answer is that George inadvertently caused a counter-revolution in the way economists, businessmen and politicians would view the process and essence of growth. How? The answer is a story of intrigue, deception and political economy marking the darkest decades of economics. In fact, some readers may have a hard time believing it. I've had a hard time believing it too! Yet there is no better explanation for the dramatic transformation of economic growth theory in the early decades of the 20th century.

Recall the phrase, "land, labor and capital." These were the factors of production long recognized by all the classical economists from Smith to Marx. They argued over which was most important, and especially whether labor or capital added the most value in the production process, but no one ever doubted that each of the three were essential. For that matter, it didn't take a classical economist to identify the three factors. From staid philosophers like David Hume to wild anarchists like Mikhail Bakunin—and especially farmers in the fields—no one could have seen otherwise. Yet, as a backlash against Henry George, *Progress and Poverty* and the single-tax movement, land was dropped from the equation. Production would now be a function of capital and labor, period.

Once in awhile a book appears from a little-known corner of academia and creeps around colleges, libraries and conferences until it gets discovered by those who find it unexpectedly relevant and important. If the discovery is sufficiently promoted, the book may catch a second wind, sailing into unexpected seas, making an impact far beyond the original printing. This will be the case, we should hope, with *The Corruption of Economics* by Mason Gaffney and Fred Harrison. Published in 1994, *The Corruption of Economics* is in some ways a very strange book. It was published by the tiny British publisher Shepheard-Walwyn Ltd., "in association," the title page tells us, "with Centre for Incentive Taxation Ltd." It is part of the "Georgist Paradigm Series," the editor of which is Fred Harrison, the director of the Centre for Incentive Taxation. Although Harrison is formally a co-author of *Corruption of Economics*, this is really a book by Mason Gaffney, born in 1923, a professor of economics at the University of California-Riverside since 1978. Harrison's contributions include a prologue and a concluding section; Gaffney penned the guts. Just to confuse the issue, one Kris Feder, an assistant professor of economics at Bard College, authored (along with Harrison) a postscript, "South Africa 1994: Countdown to Disaster." Her name appears in "About the Authors," but not the title page.

The Corruption of Economics is a somewhat roughly-bound book with an obnoxious purple cover. It includes an erratum sticker on page 23 telling us that "a Welfare Stateurce" should be "as a source." Dr. Gaffney strikes a smile at us from the back cover, but his countenance exudes intensity and drive. Gaffney is determined to deliver a message, and the back cover gives readers a sense of what they are in for:

> "To stop Henry George the fortune hunters hired professors to corrupt economics and halt democratic dialogue. The use

of that corrupted economics continues to this day, explain the authors, who analyze attempts to intimidate reforming politicians like Nelson Mandela."[6]

Despite the crude appearance of the book and its conspiracy theory overtones, Dr. Gaffney's scholarship is indisputably impressive. Gaffney is an expert on taxation and public finance, but he is especially an economic historian. In *Corruption* he focuses on post-George political economy and the associated development of economic thought, and his work is thoroughly documented. His section (pages 29–164) is entitled "Neoclassical Economics as a Stratagem Against Henry George." Blow by blow, he reveals how a select group of American land barons established the dominant economics schools and departments in the United States, populating them with faculty who were anti-George. It appears that one of the primary weapons in the war against George was the production function.[7]

A production function is an extremely simple expression that identifies what is required to produce goods and services. For a national economy, it takes the form:

$$Y = f\{x, y, z \ldots\}$$

Y refers to the sum total of goods and services produced, f is shorthand for "a function of," and x, y and z identify the factors that determine the sum of goods and services. So the production function tells us that production (and therefore income and expenditure) is a function of how much x, y and z is put into production. In theory, there could be additional factors, as "…" suggests. But in classical economics, the Big Three of land, labor and capital pretty much covered the subject, with no ellipses necessary. The classical economists recognized that many other things influenced rates of production—social conditions, religion the weather and so on—but land, labor and capital were *the* factors per se.

While the pre-classical physiocrats had identified agriculture as the sole source of production and therefore land as the ultimate factor, the neoclassical economists at the dawn of 20th-century America swung to the opposite pole and claimed land was largely *irrelevant* to the production process. Real production, they claimed, came from labor and especially capital. As much as anything, this justifies the label "American apologists."[8] To the American apologists, only taxes on wages, and to a lesser extent on capital gains, were appropriate. The physiocrats' notion of taxing only land was portrayed as ancient history. In fact, they said, land didn't even belong in the production function. "Land, labor and capital" became "labor and capital."

As noted earlier, various academic developments are used to demarcate the transition from classical to neoclassical economics. I believe this new outlook on land is by far the most distinctive and important development for our purposes.[9] I have sought other explanations for why land is not in the production function in typical textbooks. All pale in explanatory value to the anti-George backlash documented by Gaffney. In fact, only one bears mentioning. I owe this alternative explanation to an economist (whose name I have forgotten) who commented after a talk I gave at Purdue University in 2003. She pointed out that it is impossible to show more than two factors in a textbook graph. If only two factors are used, they may be placed on two offset X-axes, with production on the Y-axis, and the resulting graph takes a conical shape that illustrates how much production can be expected from various combinations of the two factors. With three or more factors, a production function becomes impossible to represent in two dimensions graphically, as in a textbook, thus making the relationships among factors more difficult for the student to envision.[10] However, none of this explains why the two factors selected are invariably labor and capital instead of land and labor or land and capital. Furthermore, if this was the answer to why only two factors were used in the production function, one would expect the textbook to explain precisely that and to clarify that, in fact, there is one more factor called "land." The typical textbook, however, does no such thing. There must be another explanation for the near-total ignoring of land in macroeconomic growth theory and instruction. Gaffney's thesis is the only compelling explanation.

Gaffney's thesis is absolutely devastating for neoclassical economics. It shows that neoclassical economics, American-style, was borne of deceit. One of the primary conceptual "advances" in neoclassical economics—the re-tooling of the production function—was nothing more than a ploy to protect the land barons from tax reforms. For Gaffney, his co-authors and Georgists worldwide, the major implication is that the single tax on land is still the most appropriate approach to economic justice. For you and I and the grandkids, however, there is an even more important implication: the concept of economic growth as described by neoclassical economics was corrupted from the start, an academic deal with the devil. This greatly helps to explain how later theories of economic growth, such as endogenous growth theory (next chapter) ended up so far adrift of the natural sciences—and from common sense. It also helps to explain why economic growth theory has become so ecologically ignorant and economic policy so environmentally damaging.

Unfortunately, *The Corruption of Economics* has not taken the economics profession by storm. That's not how it works in today's publishing world. For the little guy to strike it big, it works more like this: A wealthy interest wants a message to be delivered far and wide and assists in the delivery. Deals are cut with large publishers, front-table displays are bought at the bookstores, promotional websites are designed and book reviews are brokered into prominent newspapers. This is how, for example, a book like *The Skeptical Environmentalist*, a fallacious but apparently damning critique of the environmental movement, becomes a best-seller. It obtained the imprimatur of the Competitive Enterprise Institute and its author, Bjorn Lomborg, was suddenly seated in a high post in the Danish government. Such books are often shoddy,[11] but few are investigated. Even if they are found in violation of scientific ethics (as was *The Skeptical Environmentalist*), the damage has typically been done, and there is no counterpart to the Competitive Enterprise Institute to spread the news of the shoddiness.

Meanwhile, books that support reform agendas will be supported in spirit by other groups, but these other groups don't have the resources to get the books into the bookstores. The spirit is willing but the flesh is weak. This is a tragic barrier for truth in the corporate world of non-fiction. It also helps to explain how easy it is to be a corporately backed "conservative" (as opposed to a real *conserv*ative, such as a natural resources conservationist) in the political arena, even with the IQ of a boot. Such "conservatives" have the money (which is what makes them want to "conserve" current policies) and therefore control much of the book circulation, along with magazines, commercial programming, and other media that shape public consciousness. Alternatively, a financially struggling author can strike it rich by writing persuasive "conservatism," as such material will put her in the loop with Big Money. *The Corruption of Economics* doesn't put Gaffney in the loop, to put it mildly.

Let us take a closer look at Gaffney's thesis, just enough to give us a flavor of the evidence. Gaffney begins by lobbing this into the lap of the economics profession:

> Neoclassical economics is the idiom of most economic discourse today. It is the paradigm that bends the twigs of young minds. Then it confines the florescence [sic] of older ones, like chicken-wire shaping a topiary. It took form about a hundred years ago, when Henry George and his reform proposals were a clear and present political danger and challenge to the landed and intellectual establishments of the world. Few people realize to what degree the founders of neoclassical economics changed the discipline for the express purpose of deflecting George and frustrating future students seeking to follow his arguments.[12]

Gaffney, of course, is about to correct this lack of realization. He describes just how influential George and the single-tax movement became, especially in the United States. He asks, "Are we imputing too much weight to a minor figure? We are told that Georgism withered away quietly with its founder in 1897. That, however, is warped history."[13] He backs up his claim with a virtual laundry list of movements, policies and political parties that were influenced by Georgist thinking to varying extents during the early decades of the 20th century. The single-tax movement and a single-tax party were head-on Georgism, but Georgist political economy was also melded into the Populist Movement, the Progressive Movement and the original federal income tax law. The origins of the referendum movement, Upton Sinclair's near-governorship of California, and William J. Wallace's US presidential candidacy in 1924: all a function of George's teachings. It is reassuring to find that the attribution of these and many other important developments is not a matter of Gaffney's own musing, for he cites many other scholars who lived closer to the influence. For example, he quotes the historian Eric Goldman who, writing about *Progress and Poverty* in 1956, said, "no other book came anywhere near comparable influence."[14]

Gaffney briefly summarizes George's teaching itself and what it would mean for American society if it was followed. Perhaps the most relevant point for our purposes is that a single tax on land would strongly discourage the land speculation and urban sprawl that plagues the American environment today. The single tax would keep the agricultural sectors in the most productive lands and manufacturing and services in the most efficient locations.

We get to the "guts of the guts" beginning with Gaffney's chapter entitled "The Empire Strikes Back." This is the part of the book that will shock readers, and the voltage will be one click higher for economists, because they will recognize how big the players are in Gaffney's thesis: John Bates Clark, Edwin R. A. Seligman, Richard T. Ely, Alvin S. Johnson, Frank Knight and more. Economists will also be surprised to hear of the supportive forces who set the stage for George, most notably John Stuart Mill, Hermann Heinrich Gossen and Léon Walras. In Gaffney's straightforwardness we sense the ghost of George himself:

> As to the academic clerisy, George first suspected, and then impugned their motives. They were myrmidons of the rent-

takers, using smoke and mirrors to addle, baffle, boggle, and dazzle the laity. He provoked, supplying motive for venomous reaction from those whom the shoe fits. The inevitable counterattack came to be called "neoclassical economics"... It was a radical paradigm shift. The task was to vandalize the stage Mill had set for George, torch the old furnishings, and reset the stage permanently in ways to discomfit George and frustrate future Georgists.[15]

Gaffney begins his analysis of neoclassical motives with John Bates Clark (1847–1938). Economists know Clark as one of the fathers of neoclassical economics who helped to consolidate and refine the work of the neoclassical grandfathers (Walras, Jevons and Menger). Gaffney found 24 publications by Clark that were directed against George over a period of 28 years. For example, in Clark's review of Alfred Marshall's *Principles of Economics*, one of the leading textbooks of all time, he spent 26 pages attacking George's concept of land rent. It is by no means rare for a reviewer to use the opportunity to pick an old bone here and there, but Clark was going after the whole skeleton. His critique of George was not only exhaustive but strained. Clark aimed:

> to undercut Henry George's attack on landed property by erasing the classical distinction between land and capital. His method was to endow capital with a Platonic essence, a deathless soul transcending and surviving its material carcass. Some characterize Clark's concept as "jelly capital", some as "plastic", some as "putty", but those concrete images rather trivialize the abstract, even spiritual element, and the power of mystical traditions he could marshal to support it. There was an element of reincarnation, evoking Hinduism, transcendentalism, and Rosicrucianism. Clark even uses "transmutation", evoking alchemy. Capital was an immaterial essence, a spiritual thing, that flowed from object to object... becoming land itself. That is the only apparent reason for the mysticism, smoke and mirrors.[16]

Clark's move to Columbia University in 1895 makes the eyebrows of hindsight wrinkle with suspicion. Before receiving his distinguished position at Columbia, he was affiliated with small colleges such as Carleton, Amherst and Smith. Meanwhile, the president of Columbia was Seth Low, a wealthy silk importer and landowner who in 1895 was preparing to run for mayor of New York—against Henry George. The hiring of Clark by a wealthy landowner and political opponent of Henry George wasn't necessarily unseemly, nor would it necessarily imply that Clark's work would be swayed by his new university president. After all, Clark had already debated George in 1890 at Saratoga, New York, when he argued that capital "transmigrates" into land, shrouding the distinction between land and capital and giving the impression that capital was the "spirit" of production. In this view, land was like a lump of clay waiting for the life-breath of capital.

It does seem like more than a coincidence, however, that Clark was in high demand among other leading, anti-Georgist universities, including Johns Hopkins, the University of Chicago and Stanford. Johns Hopkins had been recently founded with Baltimore & Ohio Railroad money, the University of Chicago was the offspring of John D. Rockefeller the oil magnate, and Stanford had roots in the Southern Pacific Railroad. Rockefeller and the railroads—huge landowners—were some of the most natural enemies of Henry George, but Columbia outbid the others.

The move to Columbia allowed Clark to team up with Edwin R. A. Seligman (1861–1939), who had been Clark's ally at the Saratoga debate with George. Seligman was from a wealthy banking family and became chairman of the Economics Department at Columbia under Seth Low and then under the new president of Columbia, Nicholas Murray Butler. Butler was known for his close ties with J. P. Morgan and Wall Street, bringing money into the university and especially into the Economics Department. Columbia became the wealthiest university of the time, and the Economics Department went from two faculty members to more than forty during the Butler/Seligman administration.

FIGURE 4.1. Henry George, author of *Progress and Poverty*, and John Bates Clark, early American practitioner of neoclassical economics. George and Clark debated the nature of land as a factor of production, with massive amounts of wealth at stake. Credits: *(left)* Robert Schalkenbach; *(right)* Columbia University

The team of Clark and Seligman, supported by a flood of faculty hired by Butler, became a powerful force attacking the flanks of the single-tax movement. While Clark denigrated the concept of land as a factor of production (along with George's proposal for a single tax on land), Seligman became one of the most influential American tax economists of all time. When it comes to the distribution of wealth, this is no minor distinction. While other fiscal and monetary policy arenas may be more important to the process of economic growth, nothing can redistribute money en masse like a federal tax system. Unfortunately, when it takes a particular set of economic concepts to support a tax system acceptable to Big Money, economic growth theory can suffer the consequences.

In their efforts to divert attention from land as a factor of production, Clark and Seligman found support from other heavyweights at the front of the neoclassical transition. Most notable was Francis Walker, the first president of the American Economic Association, president of the Massachusetts Institute of Technology (MIT) and Director of the US Census Bureau. Like Clark, Walker was an early and adamant debater of Henry George. Although he later came to temper his disagreements with George, he remained in the camp that subsumed land under the concept of capital.

George and his followers were at a major disadvantage in several regards. For one thing, both Clark and Seligman outlived George by four decades. Furthermore, after *Progress and Poverty* was published George and his followers were constantly on the front lines of political battle, their economic message diluted by the many and sundry political issues. George had little time to write economics textbooks or even articles. Finally, Clark and Seligman had Big Money backing them, with all the privileges that accompany wealth (including the proliferation of like-minded faculty). They continued to write the textbooks and the tax codes long after George perished under the strain of the 1897 New York City mayoral race.

Another major figure in Gaffney's thesis is Richard T. Ely (1854–1943), educated at Columbia University, founder of the American Economic Association in 1885, and one of the most prolific economics authors ever. Ely's name is not so strongly associated with the transition from classical to neoclassical economics. Instead, he charted his own terrain: "land economics." He wrote the seminal textbook *Outlines of Economics*, first published in 1893. Ely's rationale for attacking Henry George was broader than that of Clark and Seligman's. For starters, he was himself a highly successful land speculator. Later, when he established his Institute for Research in Land and Public Utility Economics in 1920, his major contributors were utilities, railways, building and loan associations, land companies and bankers. It takes little more than an earthworm's imagination to perceive the pressures such a network could mount against a singular tax on land.

Then there was Ely's patron, Daniel Coit Gilman. Gilman was a master of exploiting the Morrill Act of 1862. The Morrill Act granted vast areas of land to the states, which were then allowed to sell the land for purposes of establishing agricultural and engineering universities. These are known as the "land-grant" schools, and they are the bedrock of American higher education. Typically,

The University of (you name the state) is a land-grant school. The smallest state grants were 90,000 acres, and over 70 land-grant schools were established pursuant to the Morrill Act. Administering these grants became a highly complex financial endeavor, with lands sometimes being used directly for university construction, but often managed as real estate for university income. In some cases, titles were transferred to private trusts, which would then manage the land (supposedly in the full interests of the university). In other words, administering the Morrill Act in many cases was hardly distinguishable from land speculation. Successful speculation often required long periods of sitting on the land without conducting any meaningful economic activity, and it was easy to sit as long as the land wasn't taxed. A lot of land and money was at stake, and many university administrators specialized in the Morrill Act. George's single tax would have threatened this entire subculture of academic administrators and the universities they worked for. Gilman was a Morrill Act point-man at Yale, then Berkeley.

Gilman then made a habit of becoming the first president of wherever he went, including the University of California, Johns Hopkins University and the Carnegie Institute. This was no minor figure in American academia. Under his tutelage, Johns Hopkins became the first major university to specialize in graduate studies. For nearly two decades, beginning in 1876, Johns Hopkins produced nearly all the American PhDs in economics, laying the foundation for the economics profession in the United States. Eleven of these PhDs became presidents of the American Economic Association. Gilman began this paradigm parade by hiring Ely as his first economics professor, drawing him away from the University of Wisconsin. (John Bates Clark and Francis A. Walker also eventually taught under Gilman at Johns Hopkins.)

Yet Gilman's reign at Johns Hopkins was not originally of his choosing. He came there from Berkeley; Gaffney describes why:

> Enter the Henry George factor. Gilman had arrived at Hopkins because he had earlier been hounded from Berkeley in 1874–75 by a crusading populist journalist, Henry George.

George, running the San Francisco *Daily Evening Post*, smelled corruption in Gilman's administration...He also smelled elitism and improper diversion of Morrill Act ("agricultural" and "mechanical") funds to "classics and polite learning." George spoke for the Grange [rural community interests]...Together they made the Berkeley citadel too hot for Gilman, who resented it. It is true, the Establishment immediately gave him a new citadel at Hopkins, just founded by a baron of the B&O Railroad, and loaded with B&O Railroad shares. Still, it must have come as a nasty jolt when the frontier battler for vulgar farmers and mechanics followed Gilman back to his new realm and appeared on the sophisticated Eastern scene as, of all things, a major intellect. This is something Gilman, the networker and administrator, never was nor could be.[17]

The stage was set for war between Gilman and George. Ely would serve as Gilman's general on the academic front with the full support of the Johns Hopkins brigade. Squads of other academics marched in to engage George on other fronts; we have already considered Columbia with Clark and Seligman, and will briefly visit one more powerhouse of American economics. As for Gilman, this is perhaps enough detail for our purposes, because soon we need to return to the central issue of what it all means for economic growth theory and politics. Besides, people (especially economists) should read Gaffney's book for themselves. I will only add to the Gilman episode that the connections within and among the leading economics institutions—from patricians to presidents to professors—amounts to a conspiratorial classic deserving of a major movie production. In a rare understatement, Gaffney quipped, "Gilman had a long reach."

Before we run one more reel from Gaffney's theatre, however, we need to briefly consider Ely's influence on economic growth theory. Ely returned to the University of Wisconsin in 1892, where he established the field of land economics and, in 1920, the Institute for Research in Land and Public Utility Economics. Ely remained an opponent of Georgist thought and politics, but he took a different tack than John Bates Clark. He did not attempt to remove land from the language of economics as Clark did. Instead, he denied that land was fundamentally distinct from capital. While Clark's work was spread universally among the student body of economics, Ely's work became prominent in the more specialized yet substantial fields of agricultural, natural resources and his own "land economics." The result was that: "Ever since, the economics profession has been poised on the balance of wonderful ambivalence. Official Clarkian theory says there is no such thing as land, but just in case there is, it is to be studied under the guidance of Ely, founder of the AEA [American Economic Association], in a separate, watertight compartment. Ely isn't so sure there is such a thing as land either, but whatever it is, it must be treated as private property, and taxed nominally if at all."[18]

It would be self-evident to political economists, but perhaps it should be noted here that to subsume land under the concept of capital has the effect of drawing two independently powerful camps—landlords and capitalists—into a unified, overwhelming force in the politics of taxation. The only available "enemy," or alternative target for taxation, is labor. It's easy to see how the American federal income tax became almost synonymous with a payroll tax. Ely's endowments from a Who's Who of Big Money must have made for a cushy job, as long as he didn't facilitate any considerations of the single tax on land.

Finally, then, we come to the University of Chicago and its "Grand Old Man," Frank Knight (1885–1972). For non-economists, it is important to note that the largest and most influential economics institution in the United States, and probably the world, is the University of Chicago. Other institutions we have discussed, such as Johns Hopkins University, MIT and Columbia University, would also rank high on the list, especially during the transition from classical to neoclassical economics. (Gaffney also implicates other major American economics programs such as Cornell,

Stanford and to a lesser extent Princeton.) Gaffney's chapter on the University of Chicago is called "The Chicago School Poison," and the following excerpt gives the flavor of the deep networking among the anti-Georgists:

> How did Knight come to Chicago? John D. Rockefeller funded Chicago spectacularly in 1892, and started raiding other campuses by raising salaries. Rockefeller picked the first President, William Rainey Harper. Harper picked the first economist, J. Laurence Laughlin, from Andrew Dickson White's Cornell (he liked Laughlin's rigid conservative and anti-populist views). Harper drove out Veblen [Thorstein Veblen, the iconoclast who wrote *The Theory of the Leisure Class*] in 1906, then died, leaving Laughlin in charge of economics until he retired in 1916. He passed the torch to J. M. Clark, the son and collaborator of J. B. Clark. Frank Knight first came to Chicago in 1917 from Laughlin's Cornell. The apostolic succession is fairly clear from Rockefeller to Harper to Laughlin to Clark to Knight.... Chicago is still the lengthened shadow of John D. Rockefeller."[19]

Knight was an extremely influential figure in the development of neoclassical economics. He probably administered more neoclassical PhDs than anyone in history, and the Chicago School is often identified as its own category of neoclassical economics. To give credit where credit is due, Knight truly was one of the greatest American economic thinkers and achieved much of his fame by dissecting the work of other great economists and schools. Many of the modern-day critics of neoclassical economics would actually appreciate Knight's railings against the excessive use of mathematics inherited from the Marginalist Revolution. Marxists would find a lot in common with his ethical critique of capitalism, especially the concentration of capital. He was not an American apologist in the mold of John Bates Clark.

Knight was also notoriously opinionated and belligerent,

though, and on one issue he agreed vociferously with Clark. Land, to Knight, was not a factor of production, at least not without capital. Therefore, there was no meaningful distinction between land and capital. Under the influence of Knight (and Clark, Seligman and a host of other economists), land and capital became perfectly substitutable, eliminating the need to include land as a distinct factor of production. Again, this concept of land was used to oppose the single tax on land and land taxes in general. According to Knight, there was no "rent" in the Georgist sense, only interest that accrued from the investment in capital. (As noted above, a tax on such interest would then have to survive the gauntlet of allied landlords and capitalists.) Knight's complicity in this agenda culminated with his 1953 publication, "The Fallacies in the Single Tax."[20]

Knight seemed to get carried away, however, insisting that there was no primary distinction among *any* of the classical factors of production: land, capital and even labor. He based his argument on the notion that the existence of all these factors is a result of "past production." The technical details of the past production principle are beyond the scope of this book, but it is an extremely abstract and cynical concept that ignores the distinction in ownership between labor and capital. As Gaffney noted, "Knight also argues that slave-owners had just [fair] title to their slaves, because of society's sanction, and—note this well—because there was open competition for the capture of slaves."[21] Knight showed how one may pick and choose among economics concepts and ideals to construct theories and support taxation agendas that favor particular sectors. As it turns out, other scholars have pointed out at least one fallacy in Knight's "Fallacies" article:

> It would be desirable to reestablish a division of factors of production according to the conventions by which original property rights are established. Human beings belong to themselves, and inalienable human effort is classified as "labor". Everything that is produced by identifiable human effort belongs to either its producer, his heirs, or the person

to whom it was legally transferred, and is classified as "capital." Whatever is left is classified as "land." This division emphasizes that the existence of land is not the result of human effort, and that taxation of labor and capital, unlike taxation of land, is intrusion into personal property rights.[22]

Unless *The Corruption of Economics* makes a bigger splash, today's wealthy landowners will simply watch Gaffney's book die a slow death, keeping their fingers crossed to the end. If it starts splashing, however, we will see them scurry and spend money on more economic hocus-pocus in the spirit of J.D. Rockefeller, J.P. Morgan and Seth Low. Georgists should no longer feel alone, however. They have helped the rest of us understand how neoclassical economic growth theory became so inane. In turn, we can help them challenge the inanity on another front, the ecological front (Chapter 6). In fact, this collaboration between ecological and Georgist economics has already begun in venues such as conferences, journals and professional society activities.

Gaffney's thesis focuses on the corruption of neoclassical economics in the American tradition. To be fair, however, some of the key figures in the transition from classical to neoclassical economics cannot be accused of such corruption. None of them served as a better spokesman for the purer side of neoclassical economics than Alfred Marshall (1842–1924). Marshall may have been petty and arrogant,[23] but he is also recognized by historians as one of the five greatest economists in history along with Smith, Mill, Marx and John Maynard Keynes (Marshall's star pupil). Marshall's magnum opus, *Principles of Economics*, was published in 1890. It was the most influential economics textbook of its era and went through eight editions in Marshall's lifetime.

Marshall is best known for his contributions to microeconomics. He compiled the principles generated by the early marginalists and added many original, insightful contributions. In a sense, he was the John Stuart Mill of microeconomics. Mill had synthesized the political economy of Smith, Malthus and Ricardo; Marshall synthesized the microeconomics of Menger, Walras and Jevons. Marshall was a well-trained and gifted mathematician, which was necessary for developing marginalist theory. It is interesting and ironic, then, that Marshall was a critic of using mathematics for purposes of developing economic theory and for communicating economic principles. He realized that an obsession with mathematics could lead economics away from reality, turning economics into an ivory-tower exercise with little relevance to society.

Furthermore, Marshall was not interested in microeconomics only. He had much to say about economic growth, and this sums it up:

> The gross real income of a country depends on (i) the number and average efficiency of the workers in it, (ii) the amount of its accumulated wealth, (iii) the extent, richness, and convenience of situation of its natural resources, (iv) the state of the arts of production, [and] (v) the state of public security and the assurance to industry and capital of the fruits of labor and abstinence...[24]

In other words, Marshall was no apologist for capitalism, and no apologist for the American apologists. He recognized that land, labor and capital were still the primary factors of production. He recognized as well the importance of technological progress and sound governance. If only he had focused more on macroeconomics, perhaps the Cambridge tradition would have had more influence than the American apologists in the development of economic growth theory and politics. The result would have surely been more ecologically prudent and more conducive to sustainability.

As with Mill and the classical economists, Marshall recognized the relevance of many academic disciplines to economic growth. As historians have pointed out, during the transition to neoclassical economics there seemed to be a sort of "physics envy" that prompted economists to apply the mathematical rigor of Newtonian physics to economics.[25] Marshall thought this was nonsense and recognized that Darwin's theory of evolution—and biology in general—had more to offer economics than did physics. In the preface to the eighth edition of *Principles of Economics*, after decades of study and hindsight, he proclaimed, "The Mecca of the economist lies in economic biology rather than in economic dynamics."[26]

Rooted in a more realistic, uncorrupted understanding of economic growth, Marshall was one of the last great economists who recognized that there were limits to growth. While he is not remembered as a growth theorist, he had a sophisticated perspective on growth. His writings reveal he was fully aware of many of the principles touted by today's growth theorists who have achieved fame for "inventing" concepts such as "human capital" (to be discussed in Chapter 5). While these recent economists have led pupils, publics and policy makers to believe in perpetual growth, Marshall never forgot the relevance of land as a factor of production, and the limits imposed by that factor. Rostow summarized his views:

> As for the long-term prospect for growth in the world economy as a whole, Marshall was…something of a limits-to-growth pessimist. Looking back and projecting forward the prospects for increase in population and the demands for food and raw materials, his sense was that, despite the potentialities of science, diminishing returns to natural resources would constrain the expansion of the world economy…and he concluded that this constraint would prevail before the end of the twenty-first century.[27]

It is a revealing sign of the pro-growth times that Rostow, a neoclassical growth theorist and American presidential advisor during the "Great Prosperity" of the 1960s, branded the brilliant Marshall a "pessimist" for recognizing limits to economic growth.

Marshall was also known for his internal struggle on the issue of socialism. Like Mill, his philosophy was strongly linked to his recognition of land as a factor of production and the resulting limits to economic growth. He was an original, true conservative in the sense that he was for the conservation of natural resources and opposed to extravagant consumption. He seemed to wear his philosophy on his sleeve: he looked more like a mountain man (albeit a somewhat sickly one) than a Cambridge man. He also recognized that, as humans pushed against the limits to economic growth, they would be pushing themselves toward a more socialistic government, losing the freedom to do whatever they wished. (The mountain men figured that out too!) He was no Marxist, however, bent on a communist revolution. Instead, he empathized with George's view of the landlord as exploiter. His brand of socialism, then, would have been more Georgist and focused on land, not capital. In 1883 he even expressed the opinion that land should be socialized in about 100 years.[28]

We have seen how the greatest of economists have been big thinkers, covering a vast terrain of academic thought, tying it all to historic events and considering long-run prospects. They have not been afraid of making predictions, either, though seldom have they specified precise time frames. It behooves us, therefore, to note closely Marshall's thoughts on when the limits to growth would be encountered. He was well aware of the premature pronouncements of Malthus and Marx. He had the historical hindsight of a century and a half of industrial economy. He had a good grasp of economic geography and compared the crowded British conditions with the wide open spaces of the American West. He recognized that some parts of the world were much fuller than others, and he knew the general rate at which economies had grown and were likely to grow in the future. He had a mathematical mind keenly capable of extrapolation. He was famously unbiased. If it were anyone else, identifying the end of the 21st century as the likely time limit for economic growth could be cast off as what scientists sometimes call a WAG (wild-ass guess). This, however, was Alfred Marshall, and the grandkids will wish he had lived a century later. Then he could have enlightened their grandfathers at a more crucial time in the history of economic growth.

We have noted that history has a profound influence on the study of economics, especially compared to the natural sciences such as physics and biology. History changes the context for economics, sometimes dramatically, creating new issues and calling for new methods of analysis. During the first four decades of the 20th century history would severely jolt neoclassical economics, the vestiges of classical economics, and Marxist economics. World Wars I and II would be fought with the Great Depression in between. All three of these events brought the focus back to the performance of national economies. Laissez faire was left in the dust because only national governments were big enough and organized enough to build war machines. And of course the Soviet Union was to emerge from the ashes surrounding Stalingrad, precipitating a Cold War with GDP on the scoreboard. While neoclassical economists were engrossed with microeconomics, history demanded macroeconomics.

War machines require vast sums of money and centralized military planning. As in the pre-capitalist days of mercantilism, nations became preoccupied not only with generating wealth but with stocking war chests. For capitalist, fascist and communist governments alike, the need arose to account for national economic production. National income accounting was born, with Americans and British leading the way. In the United States, the Bureau of Economic Analysis began calculating GDP in 1929.

The best minds in economics were employed in the service of national governments to provide recommendations on war planning, reparations and economic growth in the years between the wars and afterward. Far and away the most famous and influential of these economists was John Maynard Keynes (1883–1946), Marshall's student at Cambridge. The combined influence of the teacher Marshall and the pupil Keynes would be hard to exaggerate. In economics, the closest thing to Christ and St. Paul, or Plato and Aristotle, was Marshall and Keynes. Of the latter two, however, Keynes would have the bigger impact.

When World War I was settled with the Treaty of Versailles,

Keynes saw the writing on the wall, and it was bloody red. The treaty simply set the stage for World War II. The reparations required of Germany could not be met, and frustrated Germans would surely revolt. Keynes let this be known in *The Economic Consequences of the Peace* (1919), but politicians didn't get the take-home point. Adolph Hitler would prove Keynes correct, and thereafter politicians listened carefully to Keynes.

Keynes was also the primary consultant on economic affairs during the Great Depression. He developed the concepts that made sense of the Depression and turned Say's Law into a rule of thumb instead of a law, a rule that could be bent like a thumb for significant periods of time. Production did not necessarily bring consumption. Under certain scenarios, said Keynes, consumers could lose faith in the economy, whereby their "propensity to consume" would decline, leaving producers holding a bag of unwanted goods.

Worse yet, consumers could lose faith in the financial system, especially after an experience like the stock market crash of 1929. Banks were no longer trusted. A low propensity to consume coupled with a fear of investment caused people to hoard their money under the proverbial mattress. At best, they might buy gold as a hedge against inflation. This type of behavior would bring down industry and banks alike, and this was precisely the type of behavior that characterized the Depression, during which every worsening sign resulted in even less consumption and investment. Worst of all, the depressed business climate meant that people lost their jobs, leaving them no money to spend even if they wanted to. There seemed to be no way out.

Keynes came up with the only viable solution: governments would have to put the people back to work, providing them with an income to spend. Governments would also have to play an active role in reforming the financial system and restoring confidence in the banks and stock markets. Governments would have to provide insurance to the investor, security to the retired and welfare to the unemployed, going deeply into debt if necessary. The debt would, supposedly, be temporary, paid off after a period of economic growth.

Keynes laid out his principles and policy recommendations in *The General Theory of Employment, Interest, and Money* (1936). While other books (such as Marshall's *Principles of Economics*) taught more students, *The General Theory* was one the three most influential books on economic policy ever written, along with *Wealth of Nations* and *Das Kapital*. It is not an easy book to read, not even for economists, yet the central ideas are quite simple when distilled from Keynes's dense delivery. By the time it was published, there were plenty of applied economists in government to do the distilling, and policy makers were eager for the distillation. As with *Wealth of Nations* and *Kapital*, *The General Theory* came at a time of painful change that required a new approach. Keynes's recommendations were adopted en masse by the American President Franklin D. Roosevelt, who called his Keynesian program the New Deal. Laissez faire was dead, though later statesmen would attempt to revive it.

Keynes's approach was so radically different that it spawned a whole new school of thought called Keynesian economics. The Marginalist Revolution had brought economics out of the classical era, but Keynes was no neoclassical economist. Keynes reflected on his defection from conventional economics in the preface to *The General Theory*:

> The composition of this book has been for the author a long struggle of escape, and so must the reading of it be for most readers if the author's assault upon them is to be successful—a struggle of escape from habitual modes of thought and expression. The ideas which are here expressed so laboriously are extremely simple and should be obvious. The difficulty lies, not in the new ideas, but in escaping from the old ones, which ramify, for those brought up as most of us have been, into every corner of our minds.[29]

While far less flame-throwing in tone, Keynes's reflection should remind us of Mason Gaffney. Keynes did not address the corruption of neoclassical economics, and perhaps wouldn't have known much about it without an investigation like Gaffney's. However, like Gaffney, Keynes was particularly disappointed in the shortage of reality in mainstream economics. He thought the assumptions of mainstream economic theory (which he still called "classical" economics) "happen not to be those of the economic society in which we actually live, with the result that its teaching is dangerous if we attempt to apply it to the facts of experience."[30]

The "Keynesian revolution," as economic historians call it, resulted in the most sweeping shift in economic policy in the history of capitalist economies, which had been ushered in by Adam Smith and ushered out (in some parts of the world) by Karl Marx. However, the Keynesian revolution was not so revolutionary from the standpoint of economic growth theory. In fact, in some ways Keynes's concept of economic growth was a throwback to the boundary between physiocracy and classical economics, because he subscribed to a labor theory of value while acknowledging the essential role of natural resources. Keynes never mentioned John Bates Clark or Edwin R. A. Seligman in *The General Theory* and only made passing reference to Frank Knight, but he clearly did not agree with their crowning of capital as the primary factor of production. In fact, Keynes expressed doubt that capital should be considered a factor of production at all. In his "Sundry Observations on the Nature of Capital" he wrote:

> It is much preferable to speak of capital as having a yield over the course of its life in excess of its original cost, than as being productive.... I sympathise, therefore, with the pre-classical doctrine [that is, physiocracy] that everything is produced by labour, aided by what used to be called art and is now called technique, by natural resources which are free or cost a rent according to their scarcity or abundance, and

by the results of past labour, embodied in assets, which also command a price according to their scarcity or abundance.[31]

If it had been up to Keynes, the factors of production may have changed from "land, labor, and capital" to "labor and land," not to "labor and capital." One wonders how neoclassical economic growth theory would have evolved if Keynes hadn't been preoccupied with two world wars and the Depression. One suspects he would have fought epic battles in academia with the likes of Clark, Seligman and Knight. One has little doubt he would have prevailed. Instead, Keynes's mind was spent on principles and policies that would engender economic growth, in one case to stave off fascism, in another to alleviate a Great Depression. His legacy includes the World Bank and the International Monetary Fund,[32] even though his own propensity was toward minimizing "economic entanglement between nations."[33] We chalk up another unfortunate coincidence of history, and an ironic one at that.

There is something about Keynes that often escapes notice and is too little appreciated. This all-time great economist was never enrolled in an economics program! Apparently, he took only one course in economics.[34] Nor did he receive a PhD in any subject. He referred to formal education as "the inculcation of the incomprehensible into the indifferent by the incompetent." He did have a bachelor's degree in mathematics and later honed in on probability theory and was generally fond of math, logic, history and the arts. His lack of formal training in economics combined with his unparalleled achievements in economic theory is an extremely important lesson, especially with economic growth at the crossroads. The fact that he was not indoctrinated with neoclassical economics gave him a "clean slate" from which to build *The General Theory*. He felt that a legitimate understanding of economic affairs required a synthesis of logic, intuition and worldly experience. Perhaps his brilliance would best be summarized as a mastery of common sense.

The world wars, the Great Depression and the Cold War resulted in a renewed interest in economic growth, with Keynes at the forefront. By some accounts, Keynesian economics was a rigorous revival of classical macroeconomics, which had been eclipsed by the neoclassical focus on microeconomics. In any event, the study of macroeconomics was assisted by the increasing rigor of national income accounting, which provided a new source of data for the analysis of economic growth. One of Keynes's students, Sir Roy Harrod, took the lead in developing a theory of economic growth.

In many ways, Roy Harrod (1900–1978) was of the classical tradition. He was born in Norfolk, England, and his early studies encompassed classical literature, ancient history and philosophy at Oxford. He took this big-picture background and applied it to the dramatic, real-world problems of political economy and macroeconomics. He studied under Keynes and devoted most of his attention to the pressing issue of post-Depression economic growth. He developed a model that was to become the foundation of economic growth theory. In 1939, Harrod presented this model in a famous paper in the *Economic Journal* called "An Essay in Dynamic Theory." Later in his career he would enter politics and serve as an advisor to Prime Minister Harold MacMillan. He was knighted in 1959.

For the sake of historical accuracy and to give credit where credit is due, the Russian-American Evsey Domar (1914–1997) developed a model of economic growth simultaneously and independently of Harrod that is much the same as Harrod's model. Therefore, the foundation for modern economic growth theory is typically referred to as the "Harrod-Domar model."

Harrod's "Essay in Dynamic Theory" makes *The General Theory* look like light reading. It is not overly burdened with mathematics (nor is *The General Theory*) but requires substantial concentration. Harrod invented his own jargon throughout the essay, and one cannot jump into the middle of it—much less to the conclusion—and expect to make any sense of it. I quote the following passage as an example, not expecting readers (even most economists) to understand it but rather to reveal the abstract nature of economic growth theory, even at this early stage in its development:

> Suppose an increase in the propensity to save, which is expressed by *s*. This necessarily involves, *ceteris paribus*, a higher rate of warranted growth. But if the actual growth was previously equal to the warranted growth, the immediate effect is to raise the warranted rate above the actual rate. This state of affairs sets up a depressing influence which will drag the actual rate progressively further below the warranted rate. In numerous cases we shall have occasion to observe that the movement of a dynamic determinant has an opposite effect on the warranted path of growth to that which it has on its actual path. How different from the order of events in static theory![35]

I shall attempt a basic overview of Harrod's essay, simplifying it as much as possible without losing the essence. If even this overview is difficult to follow, readers should not become discouraged. The modern theories presented in Chapter 5 are actually simpler to explain and understand. It is important, however, to have a general sense of Harrod's model if we are to understand how neoclassical economic growth theory developed the way it did.

Harrod identifies three rates of economic growth, one of which is simply the actual rate. The other two rates are theoretical: the "natural rate" and the "warranted rate." The natural rate is the maximum rate allowed by the increase of population, accumulation of capital and technological improvement. The warranted rate is the central concept of Harrod's model. For our purposes it will suffice to think of the warranted rate as the rate of growth of demand expected by firms, on which they base their capital investment. This is the rate of growth that keeps the economy in general equilibrium, so we see here the influence of marginalist theory. However, whereas the marginalists dealt with general equilibrium at any given point in time, Harrod's model deals with rates of growth and is therefore considered a major development in economic theory. Harrod refers to general equilibrium as "static" theory; his model is "dynamic."

Harrod's essay is most famous for concluding three things. First, economic instability—boom or bust—results when the three rates differ from each other. For example, the actual rate of growth will decline when the natural rate of growth is less than the warranted rate of growth. Second, when the rates differ, systematic pressures drive them to even greater differences. For example, if the actual rate of growth falls below the warranted rate, "a redundance of capital goods and a depressing influence will be exerted; this will cause a further divergence and a still stronger depressing influence."[36] Third, it is the normal state of affairs for the rates to differ. The bottom line is that the economy is constantly subject to self-perpetuating fluctuations that either deepen recessions or invoke inflation. As with Keynes's *General Theory*, the policy implication is that government must intervene, attempting with various policies to keep the economy growing along the "knife's edge," keeping it synchronized with the warranted and natural rates of growth. Harrod provided some examples of intervention, most notably increasing government expenditure (as in the New Deal) to stimulate the actual rate of growth if it was below the warranted rate, unless the warranted rate was above the natural rate.

All this may seem horribly complex (and dismal) to the noneconomist. Frankly, one can smell in the model a great deal of complexity born of the desire to come up with something, anything, for fighting recession on the one hand and inflation on the other. Get the three rates in line, said Harrod, and the economy will run as smoothly as possible. The problem is that two of the rates are theoretical and therefore difficult to manipulate or even ascertain in real life.

For our purposes, however, it is most important to note two things about Harrod's essay. First, as the first truly rigorous, formal, mathematical (algebraic, at least) attempt to describe macroeconomic fluctuations it became widely accepted by economists as the foundation upon which future economic growth models would be built. Second, Harrod never mentioned land, natural resources or even raw materials. Economic growth in Harrod's model depends on savings (which results in capital investment) and the productivity of capital. The productivity of capital, in turn, is a matter of technological progress. Population and the labor force are considered to be growing at a constant rate and independent of the central model. Population growth helps only to determine the "natural rate" of economic growth; the challenge is to keep the warranted and actual rates in line with the natural rate.

In other words, Harrod set another precedent, presumably a less corrupted precedent than that of the anti-Georgists, for defining economic growth primarily in terms of capital and technological progress. While Harrod briefly discussed the role of population in setting the natural rate of growth, labor and land were left out of his "fundamental equation." In this respect, Harrod abandoned the classical tradition, which recognized land, labor and capital as the factors of production.

There are three possible explanations for the complete omission of land from Harrod's essay and model. The first explanation is that Harrod was still operating in an empty world paradigm, in which land and natural resources were viewed as plentiful. If so, he may have recognized land as a factor of production, but not as a relevant limit to economic growth. Land and its natural resources were simply there for the picking, and the only things that affected the rate of picking were capital investment and technological progress. The second explanation is that Harrod had accepted the arguments of Clark, Knight and the rest of the American apologists who argued that there was no distinction between capital and land. This explanation would entail that when Harrod used the term "capital," he meant "land and capital." The third explanation is that he left land out of the equation for the sake of simplifying the analysis.

There is some support for the third explanation toward the end of Harrod's article, where he said, "This essay has only touched in the most tentative way on a small fraction of the problems, theoretical and practical, which the enunciation of a dynamic theory makes it possible to formulate."[37] Nevertheless, in elaborating somewhat upon the larger "fraction of problems" left out of his essay, he still failed to mention land or natural resources. This leads us to suspect that one of the first two explanations, or a combination thereof, is more accurate. An empty world paradigm (the first explanation) would make Harrod less scrutinizing of Clark's catch-all capital concept (the second explanation), which by then had permeated the neoclassical literature. In any event, the precedent was now firmly set, from both sides of the Atlantic, for modern economic growth theory to ignore land as a factor of production.

Today's neoclassical economists may object, insisting that Harrod acknowledged the importance of land when he defined capital as including "circulating and fixed" capital, as the latter could include land. Such an objection, however, would only serve to show how confused the neoclassical concept of land has become. For example, in the *MIT Dictionary of Modern Economics*, capital has two definitions. The first is "a factor of production produced by the economic system. Capital goods are produced goods which are used as factor inputs for further production. As such *capital can be distinguished from land and labour* which are not conventionally thought of as being themselves produced by the economic system. As a consequence of its heterogeneous nature the measurement of capital has become the source of much controversy in economic theory" [italics added].[38] The second definition is "financial assets." Clearly land and natural resources are not financial assets. These two definitions of capital, then, are not consistent with the notion of land as a form of capital, fixed or otherwise.

However, the *MIT Dictionary* also provides this definition of working (circulating) capital: "the amount of current assets which is financed from long-term sources of finance."[39] An asset, meanwhile, is defined as "an entity possessing market or exchange value, and forming part of the wealth or property of the owner." The definition goes on to say that these entities include real assets and financial assets. Real assets are "tangible resources like plant, buildings and land."[40] So land, which is supposedly distinguished from capital based upon the *MIT Dictionary's* definition of capital, is found to comprise one of the subsets of capital—working capital!

Which is it? Is land a form of capital or not? If so, is it fixed capital or working capital? Neoclassical economics cannot give a cogent answer. The shenanigans of Clark, Seligman and Knight have apparently rotted neoclassical economics all the way down to its dictionaries.

This confusion in the neoclassical ranks was reflected in David W. Pearce's review of my book, *Shoveling Fuel for a Runaway Train*. Pearce, the only neoclassical economist among the book's reviewers, provided the only negative review. He complained that I made no mention of the economics literature that "has helped to transform notions of capital stocks to include human and environmental assets."[41] No doubt this supposed shortcoming of *Shoveling Fuel* had to do with my reliance on the *MIT Dictionary*, where I found that "capital can be distinguished from land and labour which are not conventionally thought of as being themselves produced by the economic system."

Not only do the ironies never cease; they seem to get more incredible. To wit, Pearce was the editor of the *MIT Dictionary*! Hopefully in the next edition he can include something about how neoclassical economics "helped to transform notions of capital stocks to include human and environmental assets."

One wonders how Pearce and other neoclassical dictionary editors would interpret Harrod's model of economic growth. Did it include land or not? I don't think they'd have a clue. I couldn't find much of a clue either. This much is clear, though: *if* the Harrod model included land, it barely did, and only as a scarcely distinguishable subset of capital. That might have been OK in 1939, but it's not good enough in a full world.

Finally, one more comment on Harrod's model is in order. Writing on his theory of growth in the *Cambridge Journal of Economics*, Danielle Besomi points out that Harrod never intended his essay to present a model of economic growth at all. Besomi insists Harrod's essay was misinterpreted, largely because readers didn't understand it. Harrod himself claimed that, "rather than providing a model of cycles or growth, he was laying the foundations of a new, revolutionary and more fundamental mode of approach to dynamic problems.... His disclaimers, however, were disregarded by commentators."[42] Why didn't these commentators listen? For one thing, Harrod didn't have the mathematical tools to express his ideas in the language of mathematics, which was rapidly becoming the preferred mode of communication among economists. His attempt to express his ideas primarily in English, with bits of algebra, came up short. It didn't help matters that Keynes (the editor of *Economic Journal*, in which Harrod's essay was published) eliminated some of the material that would have made Harrod's intent more clear.

In any case, I am not convinced that the interpretation of Harrod's model as a foundation for modern economic growth theory was inappropriate, despite Harrod's disclaimers. After all, the essay was all about explaining the rate of economic growth (or more precisely, three rates of growth). To the extent that Besomi's argument is correct, however, it suggests that the foundation for neoclassical economic growth theory is an unintended and shaky one. Furthermore, if colleagues and future economists did make a mistake by interpreting Harrod's essay as a foundation for economic growth theory, they made a much bigger mistake in assuming that the foundation was complete regarding the factors of production.

We have come a long way in our tour through economic thought. We saw in Chapter 3 how the physiocrats identified agriculture as the sole source of economic production, with land and labor the primary factors of production. We saw how the classical economists viewed themselves as students of political economy, focusing on macroeconomic processes in combination with sweeping technical and political realities. In the midst of the Industrial Revolution, as a capitalist class was born, the classical economists adopted capital as the third factor of production. We visited the twisted transition from classical to neoclassical economics, with its microeconomic focus and its attempted isolation from political affairs. We saw how political economy became microeconomics and classical economics became neoclassical economics. We were also exposed to the irony of this transition: economics had ostensibly become apolitical while land was dropped from the production function as a political response to Henry George! We reviewed the foundation of modern economic growth theory, the Harrod-Domar model in which land and natural resources were either omitted or took the form of jelly capital; we'll never know for sure. We saw how neoclassical economics got saddled with unsound, somewhat schizophrenic concepts of capital and economic growth. It remains now to address the modern theories of economic growth spawned by the Harrod-Domar model.

CHAPTER 5

Not of This Earth

> *Natural resources originate from the mind, not the ground, and therefore are not depletable.*
> ROBERT L. BRADLEY, JR.

MODERN THEORIES of economic growth have their roots in the Harrod-Domar model, which described the process of growth in terms of capital investment and technological progress. A major stepping stone from the Harrod-Domar model to the modern theories of growth was Robert Solow's 1956 paper, "A Contribution to the Theory of Economic Growth."[1] This and further work by Solow won him a Nobel Prize in 1987.

Solow's theory of economic growth centers around two processes: production and the accumulation of capital. He developed a simple model of each process. The production model describes how inputs, most notably workers (labor) and their equipment (capital), are combined to produce goods and services. Production is extremely important to the firm, of course, because it determines how much income and profit it will make. A firm can produce more, and therefore increase its income, by investing in more capital or by hiring more labor. There is a trade-off between capital and labor, however, because the firm has a limited amount of money. Therefore, it must continually choose whether to invest in more capital or to hire more labor.

The firm decides whether to invest in capital or labor by comparing the costs and benefits of each. For example, if it is in the business of digging ditches, it will need workers (labor) to dig the ditches, and it will need shovels (capital) for the workers to use. In general, if labor is cheap and capital expensive, the firm will hire more ditch-diggers before it invests in more shovels. If labor is expensive and capital cheap, however, the firm will invest in more shovels before it hires more ditch-diggers.

More specifically, the firm must compare the additional production from hiring each new worker with the additional production from investing in each new unit of capital. As you might expect after reading Chapter 4, this additional production is called the "marginal product," so that we have a marginal product of labor and a marginal product of capital. For example, the ditch-digging firm may employ 50 diggers and own 50 shovels. If the firm hires another digger, it may produce an extra 2,000 feet of ditch per year because the 51st digger can take over for other diggers as they get tired. Therefore, the marginal product of labor (the 51st unit of labor, to be more precise) is 2,000 feet of ditch. Alternatively, the firm may invest in a 51st shovel. The 51st shovel will be in better condition than the other 50 shovels, all of which are somewhat worn down, resulting in an extra 50 feet of ditches produced per year. The marginal product of capital (the 51st unit of capital, to be more precise) is therefore 50 feet of ditch.

Marginal product is converted to monetary units such as dollars, franks or yen to compare the economic benefits of hiring labor with the economic benefits of investing in capital. For example, if the firm receives $10 per foot of ditch, the marginal product of the 51st worker is $20,000, or 2,000 feet of ditch multiplied by $10 per foot. After the firm compares the costs of labor with the costs of capital, it can make a decision that maximizes the difference between income and costs. If the marginal product of capital is higher than the marginal product of labor, in monetary terms and after adjusting for costs, the firm will invest in capital. If the marginal product of labor is higher than the marginal product of capital, the firm will hire more labor.

Solow's theory focuses on this ratio of capital to workers. We refer to this ratio as the "capital/labor ratio." The idea is to maintain the capital/labor ratio that maximizes production and therefore income. Once this ratio is achieved and maintained, the economy is said to exist in a "steady state." (This is not the same concept of the "steady state economy" introduced in the preface, but is clearly related, as we will see shortly.)

In addition to hiring more labor or investing in more capital during the process of growth, the firm must also consider the fact that capital wears out, or depreciates. To maintain the steady state, the firm must continually replace its depreciated capital. For example, if five percent of the shovels wear out each year, the ditch-digging company must replace them to maintain the capital/labor ratio that maximizes productivity.

Once all the firms in an economy are able to maintain the capital/labor ratio that maximizes productivity, economic growth becomes purely a matter of population growth. In fact, as long as the capital/labor ratio remains constant throughout the economy, even if the ratio does not maximize productivity, economic growth will occur at the same rate as population growth. (There is constant pressure, however, for the capital/labor ratio to adjust to a level—the steady state—that maximizes production.) Under this condition of a stable capital/labor ratio, if the population stays constant, GDP also stays constant. This is what we referred to as a steady state economy in the preface; an economy with stable (or mildly fluctuating) population, production, consumption, and therefore GDP.[2] The difference between Solow's concept of a steady state and the steady state economy is that Solow's steady state may not produce a stabilized (or mildly fluctuating) economy at all. The economy in Solow's steady state may grow or shrink, as long as the capital/labor ratio remains constant.

Both usages of the term "steady state" are found in the literature, so readers must take some care to identify which type of steady state they are reading about. Typically one can tell the difference by looking for the word "economy" at the end of the phrase. Solow's notion of a steady state (stable capital/labor ratio) is usually called just that; "steady state." For example, an author may say, "Transition economies move toward the steady state," or "Once the economy has reached the steady state, the capital/labor ratio remains constant." Sometimes the author may use the concept as an adjective, in which case the phrase is hyphenated as "steady-state." For example, the author may refer to "steady-state capital investment" or, more ironically if not oxymoronically, "steady-state growth!"

The steady state economy of stable GDP, on the other hand, is usually called "steady state economy," although it is sometimes referred to as "steady-state economy." Either usage is acceptable. "Steady-state" may be hyphenated because it can serve as an adjective to "economy." In other words, a "steady-state economy" is an economy in a state of stable GDP. However, even this creates potential for confusion because Solow's steady state (with growing GDP) is occasionally and unfortunately referred to as a "steady-state economy."

Fortunately, the unhyphenated "steady state economy" may be used to describe an economy of stable GDP and it is seldom used in the context of Solow's steady state. "Steady state" may be left unhyphenated because "steady" is itself an adjective for "state economy" (for example the economy of a nation). Thus, a truly steady, state economy, or "steady state economy," is a steady, non-growing and non-shrinking national economy.

I have tried to cover all the bases here, but these rules of thumb work in the vast majority of cases: 1) "Steady state," without being followed by the word "economy," usually refers to a stable capital/labor ratio, whether or not GDP is growing; 2) "Steady state economy" refers to an economy of stable or mildly fluctuating GDP. Mainstream growth economists are well-versed in the "steady state" concept, but are seldom familiar with the concept of a "steady state economy." We will not revisit the steady state economy in any depth until Chapter 6. We return now to the summary of mainstream economic growth theory.

In Solow's notion of "steady-state growth," if the population grows and the capital/labor ratio remains constant, the economy grows at the same rate as the population. This is only a slightly more sophisticated way of looking at increasing GDP based on population growth than what we saw in Chapter 2. Solow would not have won the Nobel Prize for this observation. He went a little further, however, and confronted the concept of technological progress.

In Solow's basic model, technological progress is an increase in production per unit of labor. This can happen if the workers work more efficiently or if the capital they employ is technologically improved. For example, the ditch diggers can learn to work as a team, or backhoes may replace shovels. The key point is that more goods and services are produced for the same amount of money spent on capital and labor. More income is received per unit of the firm's expenditure. This allows the economy to grow even if population remains constant. In other words, in Solow's model, technological progress allows for economic growth based entirely on per capita production and consumption. As we saw in Chapter 2, this is usually the more desirable form of economic growth because it makes the individual citizen richer without contributing to population growth, which in many countries has become worrisome. (We also went further to question how much better growth based on per capita consumption was, especially in the United States and Europe, but in mainstream economics it is assumed that economic growth based on per capita consumption is desirable, pretty much always and everywhere.)

One of the assumptions Solow had to settle for was that technological progress is "exogenous." This means that technological progress is not influenced by the actions of the firms producing goods and services. In other words, technological progress happens automatically, raining down "like manna from heaven." This, of course, is a very simplistic assumption because we know that firms do indeed strive to improve the technologies they use. Many firms spend money on research and development for this sole purpose. As we will see, more advanced models of economic growth have moved beyond assuming exogenous technological progress and employ a more sophisticated concept called endogenous technological progress.

Solow's assumption of exogenous technological progress was convenient for purposes of economic analysis, however. When economists looked at the data on increases in productivity and GDP around the world, they found that increases in labor and capital did not fully account for these increases. For example, from 1960 to 1990 US GDP grew by an average of 3.1 percent per year. Almost one percent was attributed to capital accumulation, and another 1.2 percent to growth in the labor force. Economists using Solow's theory could account for the remaining one percent by simply assigning it to the category of exogenous technological progress. In other words, technological progress accounted for almost one third of GDP growth.

When economists apply Solow's model, they typically assume that population will continue to grow. As long as it grows, firms tend to hire more labor. There is always pressure to maintain a capital/labor ratio that maximizes production, however. The firm must continually invest in more capital to keep up with its growing work force. Where does the firm get the money to invest in more capital? Partly from its profits, but it will also borrow money from lenders. Lenders, in turn, acquire the money from private investors. This brings us to another aspect of the Solow model.

Workers have a choice about what to do with their income. They can either spend it on goods and services, or they can save it for future purchases. As we all know, people who save don't just put their money under a mattress. They may bring it to a bank, which will lend it to firms, or they may invest directly in a firm by purchasing stocks. (In real life another option is investing in government bonds, but government operations are not addressed in the Solow model.) In any case, one of the fundamental principles of economics is that, in any economy, savings equals investment. This principle is incorporated in Solow's model of economic growth. The upshot is that, in nations where people save more money, more money is invested, which allows firms to acquire more capital while hiring more labor, thus maintaining the optimal capital/labor ratio as the population grows.[3]

For readers who may feel a little lost in all the details provided to this point, let us summarize the highlights of the Solow model of economic growth. An understanding of these highlights should suffice for understanding the rest of the chapter, because all current mainstream theories of economic growth are built upon the Solow model.

First, the basic factors that affect GDP growth are capital and labor. The amount of labor available depends upon the population size. Generally, as a population grows, its economy grows. Second, there is always an optimum capital/labor ratio that maximizes productivity. Productivity is a measure of how many goods and services are produced given the amount of labor and capital employed. In order to compare productivity figures among various types of firms, productivity is measured in monetary units.

The Solow model emphasizes that economies may follow two major paths. One is called the "transition path." The transition path is taken by economies that are moving toward a steady state. For example, if the ratio of capital to labor is too small for steady-state conditions, firms will invest in capital faster than they will hire additional workers. The economy will follow a transition path until the capital/labor ratio reaches the steady-state level. During this transition, the economy grows faster than the population. In other words, per capita consumption grows faster than GDP.

The second path is called the "balanced growth path." An economy follows a balanced growth path when capital, labor, consumption and output all grow at constant rates. Theoretically, this may occur even when the capital/labor ratio is not at its optimal level. For example, a communist government could maintain a higher or lower capital/labor ratio for various purposes, and balanced growth could result. All market economies, however, are guided or pressured toward steady-state growth with the optimum capital/labor ratio, so that most economies on a balanced growth path are in (or close to) a steady state. The US of the 1990s, for example, was thought to be on a balanced growth path.

If no technological progress occurs, steady-state conditions result in increasing GDP (because population is growing) but not in increasing per capita consumption. However, Solow's work went further to show that per capita output may increase even when capital and labor reach a steady state. This additional increase he assigned to technological progress and assumed it arose exogenously, or without any influence of producers. Of course, Solow did not really think producers have no influence over technological progress, but he built this assumption into his model to keep it simple and mathematically tidy.

For our purposes, one of the most important aspects of the Solow model is that, after an economy settles into a steady state, the only way for per capita consumption to increase is through technological progress. This means that technological progress determines the rate of per capita GDP growth. This has led economists to refer to technological progress—not population growth and not increasing capital stocks—as the "engine of economic growth." In the Solow model, technological progress occurs constantly and continually, meaning there is no limit to economic growth. Technological progress continues to rain down like "manna from heaven" and economic growth proceeds in lock-step, as it has for over 200 years. There is nothing built into the model, such as a limited amount of land or supply of natural resources, to suggest a limited supply of manna. Again, this is the basic model of economic growth that all mainstream theories are built upon.

I paid Solow a compliment in my first book, *Shoveling Fuel for a Runaway Train*. I am an advocate of common sense for the sake of revealing glaring errors in economic analysis, so I referred to what Solow himself called a "profound warning" on display in his office: "No amount of (apparent) statistical evidence will make a statement invulnerable to common sense."[4] I have become less confident, however, in the commonness of Solow's own sense. Despite the path-breaking aspects of his model, Solow stuck to the neoclassical tradition of neglecting land as a distinct factor of production. Solow is even credited with coining the phrase "jelly capital."[5] This almost sounds like a self-deprecating acknowledgment that the melding of land and capital into a single factor of production is a weakness of neoclassical economics. However, Deirdre McCloskey, an economist who has made a living analyzing the rhetoric used by economists, claims that Solow was a master of the art of persuasion. She cites several examples, including one in which Solow introduced a concept (the aggregate production function) in an apologetic tone, as if he himself was skeptical of its merits. According to McCloskey's analysis, however, Solow was neither apologetic nor skeptical: "he is pretending to be for rhetorical effect."[6] Having introduced the concept with such rhetoric, thus making it more acceptable to *truly* skeptical readers, he then went on to center the rest of the paper upon its merits. In any event, it is hard to mistake what Solow really meant in a 1974 article in which he stated that due to technological progress, "The world can, in effect, get along without natural resources."[7] He saw no limit to technological progress and no limits to economic growth.

Significant improvement to the Solow model (although not from an ecological standpoint) came during the 1980s and early 1990s when several economists, some working independently and others together, incorporated the concept of "human capital." These economists included Robert E. Lucas, Gregory Mankiw and David Romer, among others. Charles I. Jones provides a synthesis of their work in his textbook, *Introduction to Economic Growth*.[8] We will very briefly explore the concept and the implications for economic growth.

"Human capital" is shorthand for education and skills. A firm may invest in human capital by training its employees, for example, or by hiring more highly paid employees who already have a higher level of education. Workers, meanwhile, can accumulate more human capital for themselves by going to college or trade school. This means they have less time to work, however. Therefore, they have to weigh the economic benefits of education (which will take the form of higher income in the future) with the economic costs (lower income in the short term). Economists typically assume that a year of schooling results in an approximate ten percent increase in wages.

Economists have incorporated the concept of human capital into the Solow model by simply accounting for the skill level of the labor force. A worker with a higher skill level is more productive than a worker with a lower skill level. Therefore, a nation with a higher average skill level will have a larger GDP and higher per capita consumption, all else equal.

Of course, it doesn't require an exercise in calculus to conclude that a higher skill level in the labor force will result in higher productivity. Common sense will do, especially when supplemented by experience in the workplace. But mathematical modeling has allowed economists to assess various alternative scenarios, such as what may happen to long-term economic growth rates with different levels of human capital investment.

The findings are not much different than with the basic Solow model. Just as investing in "regular" capital makes the labor force more productive, investing in human capital does too. Meanwhile, the market continues to push the economy toward a steady state with a constant regular capital/labor ratio and a constant human capital/labor ratio. In other words, investing in capital (regular or human) will increase the rate of economic growth only until the steady state is reached. In the long run, however, these investments have no effect on economic growth rates. They affect only the level reached by per capita consumption and GDP in the steady state. The only things that affect the rate of economic growth at all times (short-term and long-term) are population growth and technological progress. Of these two, common sense would suggest that only technological progress could continually raise *per capita* production and consumption. Furthermore, unlike capital (regular or human), technological progress is not pressured toward a steady-state ratio.

Thus far, then, we have a mainstream theory of economic growth in which technological progress is the "engine." An engine is an odd metaphor, though, because technological progress is still raining down like manna from heaven, albeit with a work force more capable of using it. No limit to the rain has been perceived, and no limit to economic growth has been acknowledged.

I should mention that this is not an exaggeration. I spoke with Robert Lucas following his presidential address at the 2002 American Economic Association conference in Washington, DC. I asked him if he thought there was any limit to economic growth. Without hesitation he simply said, "No, because we have technological progress." I also asked him if he was familiar with Herman Daly's work on the steady state economy, and he thought he'd heard of it but wasn't conversant with it. Evidently the irony of his article, "Making a Miracle," was lost upon Lucas.[9] Manna from heaven is a miracle. Lucas didn't believe in miracles per se, and he knew that technological progress wasn't manna from heaven. Nevertheless, he believed in "making" miracles (such as perpetual economic growth) via technological progress. As I'll describe in Chapter 8, technological progress neither derives from miracles nor produces miracles.

Meanwhile, the next big development in mainstream economic growth theory stemmed largely from the work of Paul Romer, an economics professor in the Graduate School of Business at Stanford University. Beginning with his 1983 PhD research, Romer has been the lead developer of "new growth theory" or "endogenous growth theory." As a testimony to the importance of economic growth theory in national affairs, Romer was named one of America's 25 most influential people by *Time* magazine in 1997. However, most economists do not see Romer's work as a fundamental departure from the earlier models of economic growth. Therefore, I will refer to it as endogenous growth theory. It is, however, the most recent advance in mainstream economic growth theory and represents the state of the art.

As with the models of economic growth that incorporate human capital, endogenous growth theory is built upon the Solow model. The important distinction comes with how endogenous growth theory handles technological progress. As we have seen, in the basic Solow model, technological progress is viewed as exogenous, meaning it occurs independently of the firms using it. When

FIGURE 5.1. Robert Solow (*above left*) and Paul Romer (*above right*): economists representing the development of neoclassical growth theory. Their findings, far removed from ecological insight, served "pop economists" such as business professor Julian Simon (*below left*), and pro-growth politicians such as Jack Kemp (*below right*). Credits: (*top left*) Olaf Storbeck; (*top right*) Wikimedia Commons; (*bottom left*) University of Maryland; (*bottom right*) US Department of Housing and Urban Development

an economist uses the Solow model with exogenous technological progress, he or she simply assigns some value for the contribution of technology to economic production. For example, she may assign the figure based upon historical evidence, as we observed earlier. Once the economist predicts the rate of population growth and determines the capital/labor ratio in steady state, she may then demonstrate how economies could grow in per capita terms based upon rates of technological progress that may differ from historical rates.

Endogenous growth theory offers a more sophisticated approach to technological progress. As with the basic Solow model, it emphasizes that the level of technology influences the productivity of capital and labor. However, the level of technology is viewed as a "stock of knowledge," or ideas, that are used to increase productivity.

As with the models incorporating human capital, endogenous growth theory views the laborer as having divided interests. Recall that in models incorporating human capital, laborers have the basic choice of working or going to school. In endogenous growth theory, labor either works at producing goods and services or at producing ideas that may then be used to produce goods and services more efficiently. The labor of producing ideas is called research and development, or R&D, and it is hired by firms.

Generally, more R&D produces more ideas and therefore increases the productivity of capital and labor, resulting in increasing per capita GDP. However, one of the key questions for endogenous growth theory is whether there are diminishing returns to R&D. If, for example, the biggest ideas are produced first, with ever smaller ideas coming after, then there will likely be diminishing returns. In other words, there will be a limit to how much a firm will pay to conduct R&D, a limit to how many new ideas are produced and a limit to per capita GDP. But the economics of ideas are not as simple to understand as the economics of capital. In fact, many economists (including Romer) believe there are increasing returns to ideas because each new idea can be combined with existing ideas such that the effect on productivity continues to increase with every new dollar spent on R&D. Of course, the intermediate scenario is when there are constant returns to ideas, whereby each dollar invested in R&D results in a certain productivity increase, no matter how much is spent.

What complicates things further is that ideas, unlike capital, are hard to "own." This affects the willingness of firms to invest in R&D. In economics, there are two concepts that are especially relevant to the ownership of ideas. The first is the concept of "rivalry." If a good or a service is rivalrous, it can be consumed by one and only one person. For example, food is a rivalrous good. Your consumption of a candy bar prevents anyone else from consuming it. If a good or a service is non-rivalrous, it can be consumed by multiple people, even simultaneously. For example, when a music group sells its services to a shopping mall, all who enter the mall will enjoy the service, and the shopping mall will have little control over who enters and leaves. The music is a non-rivalrous service.

In general, goods tend to be more rivalrous than services, but there are many exceptions. For example, a painting in an art gallery is a good that is non-rivalrous because many people may enjoy it, even simultaneously. Conversely, a massage is a service that is rivalrous; only one person may receive a particular massage, although another person may receive a similar massage afterward.

Ideas are clearly non-rivalrous. Your having an idea does not prevent my having the same idea, and vice versa. On the other hand, if you are the first one to have the idea, you might be able to keep it a secret, at least until someone else has the same idea independently.

This brings us to the second concept of idea ownership, called "excludability." When a firm invests in R&D and an idea is produced, it may not be easy to exclude other firms from using the idea once they see it in action. Patent law, however, makes ideas more excludable. A patent is a legal document that describes the precise attributes of an idea or invention and entitles the holder to monopolize its use. A patent, therefore, establishes an "intellectual property right."

In the United States, a patent usually has a time limit ranging from 17 to 20 years from the filing date. If firms did not have this ability to exclude others from using their ideas, they would have far less economic incentive to conduct R&D. Economists consider intellectual property rights an extremely important requirement for technological progress and therefore economic growth. A firm that invests in R&D and produces an idea that increases the productivity of its labor and capital can patent the idea and exclude other firms from using it for at least 17–20 years. The firm with the patent is now in a nice economic position, especially if it did not have to spend very much on the R&D. The R&D costs were one-time costs, and the firm now has exclusive rights to the idea. It can be applied to as many products as consumer demand will allow, adding to the firm's profit, or "increasing returns" with each new consumer.

For example, people are always looking for the proverbial better mousetrap. If a firm invented one, there would be tremendous demand. The firm would have spent on R&D for the invention just one time, or for some period of time, yet would keep selling mousetraps over and over again until the huge demand was met. With each new mousetrap sold, the money spent on research would become a smaller fraction of the firm's expense. The more mousetraps sold, the higher the profit per mousetrap. This is what is meant by increasing returns to scale, and it opens up a whole new prospect for economic growth theory. Indeed, it is one of the most distinctive aspects of endogenous growth theory.[10]

Alas, however, as economists are fond of saying, "There's no such thing as a free lunch," and thus it is with R&D. Increasing returns are not a free lunch except in theory, and even then, only when protected by patent. Eventually, no one will need another mousetrap for a long time, and sales will slow to a crawl. The firm still has other costs, and if sales are too slow, the returns are offset by capital depreciation, land rents and operating costs. Furthermore, after the patent runs out, full competition returns and drives the profits down to nothing. An even better mousetrap must be invented; technological progress must continue if per capita GDP is to keep growing.

For today's neoclassical economist, there is no reason to believe technological progress won't continue, because we have intellectual property rights that make it worthwhile for firms to invest in R&D. True, technological progress is no longer raining down like manna from heaven, but heaven is no longer necessary, sayeth the neoclassical economist, because firms are actively engaged in technological progress.

This brings us to one more extremely important conclusion that stems from endogenous growth theory: population growth must occur for per capita GDP growth to continue in the long run.

Ahem. *Did you catch that?*

Yes, endogenous growth theory would have us believe that population growth is not only required for GDP growth to continue in the long run, but for *per capita* GDP growth in the long run.

Just to be doubly clear: if we want more income and more consumption *per person, we need more people.*

This bald and bold conclusion may shock readers with an ecological background—and many others with a background in common sense—especially after all that has been said about technological progress as the "engine of economic growth." It certainly would have shocked Quesnay, Malthus, Ricardo, Mill, George and Marshall, and probably Smith and Marx, and even Keynes. What's behind it?

As mentioned earlier, the endogenous growth model accounts for the "stock of knowledge," or the amount of ideas accumulating over time as a result of R&D. However, if the growth rate of ideas falls, so does technological progress and therefore per capita GDP. As Charles I. Jones put it in his textbook, "In order to generate growth, the number of new ideas must be expanding over time. This occurs if the number of researchers is increasing—for example, because of world population growth."[11]

Actually, Jones understated his case with the "example" of population growth, because there is *nothing* short of population growth that can keep the number of researchers (and the growth rate of ideas) increasing in the long run. If the population stopped growing, eventually everyone would have to be a researcher to keep the growth rate of ideas increasing. This would be impossible, of course, because it would leave no one to use the ideas in the production of goods and services. Even if it were possible, once the number of researchers reached 100 percent of a non-growing population, the growth rate of ideas could not increase and no more growth in per capita GDP would occur. The only way out of this conundrum—excluding manna from heaven—is if the constant stock of researchers produces ideas at an increasing rate. The researchers have a limited amount of time in a day, however, and their production of ideas can't be expected to increase indefinitely. Therefore, population growth is ultimately depended upon for the sake of not only increasing GDP, but also for the sake of increasing *per capita* GDP. In this sense, then, endogenous growth theory is indeed a departure from the Solow growth model, in which population growth may contribute to GDP growth in the long run, but not per capita GDP growth in the long run. Perhaps endogenous growth theory deserves the title of "new growth theory" after all.

If I were reading this I would be asking, "Could this Czech fellow be making a straw man of endogenous growth theory? How could something so fancy-sounding—'endogenous growth theory'—lead to something so naive? He must be exaggerating." That's what I was afraid of, too, so just to be safe I called Charles I. Jones at the University of California, Berkeley, to make sure I was reading his textbook correctly. Sure I was, "Chad" (Jones's handle at UC-Berkeley) informed me.[12]

The lessons from this investigation are two-fold. First, readers may rest assured that I have not been making a straw man of endogenous growth theory. More rigorous studies of endogenous growth theory than mine also find it fallacious. For example, Tommaso Luzzati looked at the finer nuances of a wide variety of endogenous growth theories and found, "Although endogenous growth models avoid simplistic representations of the links between the economy and the environment, the conditions for unlimited growth are built on attempts to break exactly those links, that is, on attempts to decouple matter from the economy."[13] Decoupling matter from the economy is a violation of the first law of thermodynamics, which every economist should know in its plain-language form: "You can't get something from nothing."[14]

To be fair, Romer's general sketch of things—stock of ideas, non-rivalness of ideas, patent law—is logical, even brilliant. Romer is widely known for his creativity, and he has too many admirers who think he's a genius for him to be a dummy.[15] The problem is that his genius has one fatal flaw, at least, because it leads us (or at least the gullible among us) to conclude that we can expect per capita GDP to increase, on into the long run and without end, *as long as population continues to grow.*

Like the brainless scarecrow in the Wizard of Oz, endogenous growth theory may contain a lot of "straw," but like the wizard in the final scene, we are not responsible for it. We didn't build a straw man here. The fact is that endogenous growth theory has come unhitched from the real world. It has a fatal flaw and it's having fatal consequences.

Perhaps the economic growth theorists aren't entirely responsible for the straw either. They seldom have a background in ecology and physics, leaving them somewhat gullible to fish stories about perpetual growth. On the other hand, ignorance of the law (laws of physics, in this case) is no excuse for making such wild-eyed claims that run so contrary to common sense.

PART 3

ECONIMICS FOR A FULL WORLD

CHAPTER 6

Ecological Economics Comes of Age

> *The ideas which are here expressed so laboriously are extremely simple and should be obvious. The difficulty lies, not in the new ideas, but in escaping from the old ones, which ramify, for those brought up as most of us have been, into every corner of our minds.*
>
> JOHN MAYNARD KEYNES

OF ALL THE CRITIQUES of mainstream economics—Third World, feminist, Austrian, radical, Georgist, Marxist and others—the one our grandkids would have us heed most is the ecological critique. The ecological critique says that mainstream economics has ignored some extremely important scientific principles that are especially relevant to economic growth in the 21st century. These principles, taken together, make it abundantly clear that there are limits to population growth and to the production and consumption of goods and services, no matter how efficiently we try to produce and consume. In other words, these principles make it clear that there is a limit to economic growth. Therefore, a full world in pursuit of economic growth finds itself in violation of the laws of nature and is penalized accordingly. As they say, "Nature bats last." Unfortunately, the penalties will be most severe for the grandkids, and this will be supremely unfair because the grandkids will have had no say in the formulation of our economic goals.

The ecological critique of mainstream economics is so strong and compelling that a large and growing academic movement has formed around it. This movement is called "ecological economics," no less, and is more or less embodied in the International Society for Ecological Economics, or ISEE.[1] There are ISEE chapters representing the United States, Canada, Europe, Russia, Australia and New Zealand, Brazil, Argentina, and India.

As with most movements, there are various views on how ecological economics originated. However, at least three couplings of people and their thought-provoking writings would be prominent in any discussion of ecological economics history. One is the controversial book by Donella Meadows, Dennis Meadows and Jorgen Randers called *Limits to Growth*, published in 1972. Another is the highly theoretical work of the Romanian professor Nicolas Georgescu-Roegen, summarized in his book *The Entropy Law and the Economic Process* (1971). The third would be the profound but down-to-earth work of Herman Daly on the steady state economy, featured in books such as *Valuing the Earth* (1993), *For the Common Good* (1994) and *Beyond Growth* (1997).

Limits to Growth was a cornerstone of the American environmental movement and was eventually translated into 30 languages. The authors, based at the Massachusetts Institute of Technology (MIT) and commissioned by the Club of Rome, developed a computer model demonstrating how economic growth was leading to natural resource depletion and environmental degradation. Two of the computer scenarios, including a "business as usual" scenario and a dramatic technological progress scenario, predicted a disastrous collapse of the economy during the 21st century. The third scenario was essentially the steady state economy and assumed concerted efforts to stabilize the system. The book and its authors suffered a politically debilitating attack in the decades following its publication. At first, economists in academia chipped away at details, but soon pro-growth, free-market organizations such as the Competitive Enterprise Institute and Cato Institute piled on with an overarching accusation of "pessimism." Such criticism was similar to the 19th-century criticism of Malthus's *Essay on Population* and is hard to read without countering: "Don't throw the baby out with the bathwater." Perhaps Meadows and her colleagues weren't spot-on with every detail, but the principles they laid out were undeniable and the scenarios were rigorously constructed. Decades later analysts are documenting how prescient the authors of *Limits to Growth* were, especially with the business as usual scenario.[2]

In contrast to *Limits to Growth*, Georgescu-Roegen's masterpiece went mostly unnoticed in academia and was entirely ignored in public dialog. It's effect has been like the hands of time, tick-tocking perpetual growth notions into the dustbin of yesteryear's fantasies. The slow but sure ticking is apropos, given that *The Entropy Law and the Economic Process* is all about "time's arrow," or the entropy law.

The entropy law is a foundational concept in physics: the second law of thermodynamics no less. Perhaps the quickest, easiest way to describe it is that energy inevitably, invariably dissipates. Things that are hotter than their environment cool off. Of the billions of cups of coffee poured in the broad sweep of history, not one has warmed up of its own accord, not for an instant. The entropy process is as consistent and irreversible as Father Time; you can tell whether it's earlier or later based on the warmth of your coffee. Einstein said of the entropy law, "It is the only physical theory of universal content, which I am convinced...will never be overthrown." Einstein was also impressed by the entropy law's "range of applicability."[3]

And apply it Georgescu-Roegen did, unto 457 pages! The main application, in a nutshell, is that absolute efficiency in the economic production process cannot be achieved. Nor can recycling be 100 percent efficient. Pollution is inevitable, and all else equal, more economic production means more pollution. These findings may seem like no-brainers to many, yet neoclassical growth theory has led to wild-eyed optimism regarding "green growth" and "closing the loop" by turning all waste into capital. Such fantasia cannot be soundly refuted without invoking the entropy law.

The Entropy Law and the Economic Process moves across a huge swath of philosophical and scientific terrain. As with most wide-ranging and intellectually adventurous books, *The Entropy Law* can and has been challenged. Most of its arguments and the counter-arguments are philosophical and not amenable to scientific proof or disproof. But the tremendous value of *The Entropy Law* is that it unequivocally established the profound relevance of thermodynamics to economic affairs. Unlike neoclassical economics, ecological economics embraces this relevance, putting ecological economics into a better position for enlightening real world affairs.

With regard to real world affairs, though, *The Entropy Law* as a book was not as useful as the entropy law itself. It was abstruse enough to appear esoteric, and Georgescu-Roegen's interests in economic affairs tended to be exceedingly long-term. While neoclassical economists pushed a perpetually growing economy, Georgescu-Roegen emphasized a perpetually eroding economy and indeed a perpetually eroding universe, all the way out to the "heat death" necessitated by infinity. This emphasis had the ironic effect of retarding the application of *The Entropy Law and the Economic Process* to the economic process itself.

Fortunately for ecological economics, one of Georgescu-Roegen's students at Vanderbilt University was Herman Daly. A devout Christian, Daly too had an eye toward the longest of long terms, but he also had one eye focused on the wellbeing of present and upcoming generations. This tapestry of long- and short-term interests can be sensed throughout Daly's writings. Daly took the entropy law, emphasized its short-term relevance while acknowledging its long-term implications, and used it as part of a well-grounded macroeconomic framework. He called this framework "steady state economics," which served as the catalyst for the ecological economics movement. Much of the remainder of this book is a natural progression from Daly's steady state economics.

With the passing of Georgescu-Roegen (1906–1994) and Donella Meadows (1941–2001), of the three only Daly, a professor emeritus with the University of Maryland, remains a major figure

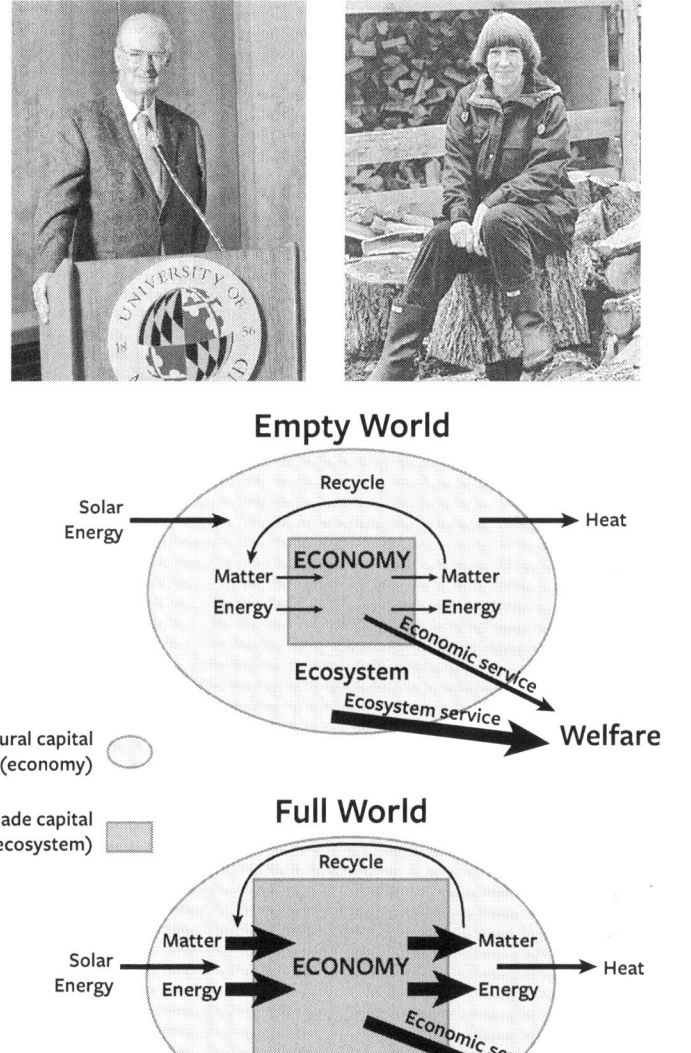

FIGURE 6.1. Herman Daly (*top left*) and Donella "Dana" Meadows (*top right*), founders of ecological economics. Daly and colleagues clarified the relationship between the economy and Earth with a diagram (*above*) that was simple but powerful for illustrating limits to growth. Credits: (*top left*) Herman Daly; (*top right*) Donella Meadows Institute; (*above*) From *Ecological Economics*, Herman E. Daly and Joshua Farley, ©2004 Herman E. Daly and Joshua Farley. Reproduced by permission of Island Press

in ecological economics.[4] *For the Common Good* (co-authored with the theologian John Cobb) received the prestigious Grawemeyer Award for Ideas Improving World Order. Daly was also the recipient of the Honorary Right Livelihood Award (Sweden's alternative to the Nobel Prize) and the Heineken Prize for Environmental Science from the Royal Netherlands Academy of Arts and Sciences. Daly is no ivory-tower academic, either, having spent six years as a senior economist at the World Bank. The National Council for Science and the Environment presented Daly with its Lifetime Achievement Award in 2010. The tremendous respect for Daly is displayed in a festschrift authored by colleagues, students and admirers.[5]

Of course, it is somewhat arbitrary to classify these relatively recent efforts as the "roots" of ecological economics. We saw in Chapter 3 that the classical economists, most notably Malthus, recognized limits to economic growth. John Stuart Mill went further and elaborated on the "stationary state." Daly's steady state economy is essentially the resurrection of Mill's stationary state, supplemented with a rigor gleaned from the natural sciences, an economic mastery honed in academia and first-hand experience with economic growth policy as implemented by the World Bank. After six years at the Bank, Daly left in disgust, noting the blind faith in neoclassical economics among the Bank's highest-ranking economists.[6] He offered the hopeful observation, however, that the Bank was becoming "more environmentally sensitive and literate."[7]

Daly was modest, for the newfound environmental sensitivity wasn't foisted onto the Bank by Wall Street or the Competitive Enterprise Institute. In fact, we have the likes of Daly himself and a noteworthy colleague at the bank, Robert Goodland, to thank.

With that brief historical account as background, the remainder of this chapter will comprise an overview of ecological economics, with an emphasis on how ecological economics treats the subject of economic growth. As with conventional economics, ecological economics can be broken down into micro- and macroeconomics. Ecological economics is founded upon different principles, micro and macro, which lead to distinct conclusions and policy implications. These principles stem from the natural sciences (physical and biological) that are largely ignored in conventional or neoclassical economics.

While we keep the micro-macro distinction in mind, it will also be useful to think of three themes: allocation, distribution and scale. "Allocation" refers to the way the factors of production are devoted to different producers for different purposes. For example, land may be allocated among farming, forestry, recreational and other uses. Labor and capital may be allocated likewise. At a finer level, timber from a forest may be allocated among furniture-making, construction, boat-building, etc. Labor at the construction site may be allocated among carpentry, masonry and plumbing. Capital at an automobile plant may be allocated among the chassis, drive-train and circuitry floors. The efficiency of an economy depends to a great extent upon a well-balanced allocation among and within the factors of production.

"Distribution" refers to the distribution of income, wealth or general welfare. This is the economic subject most often discussed by non-economists. Indeed, politics is mostly about distribution, which explains the classic definition of politics: "Who gets what, when, and how."[8] Bill Clinton could have elaborated, "It's the political economy, stupid!"

"Scale" refers to the size of the human economy relative to the ecosystem. This, of course, is our focus here, and it provides the primary distinction between neoclassical and ecological economics. Neoclassical economics deals almost exclusively with allocation and, to a much lesser extent, distribution. Why? Because neoclassical economics doesn't recognize environmental limits to economic growth. With no limit to growth, the concept of scale is superfluous, there is no conflict between growth and the environment, and the cure for social ills—including maldistribution of wealth—is always more growth. "A rising tide lifts all boats," as they say.

Ecological economics deals with allocation and distribution, but its emphasis is on scale, especially among the scholars and policy activists we might call "Dalyists." Scale deals with whole economies, usually national or global, so ecological economics is geared especially to replace conventional *macroeconomics* while accepting and incorporating some of the fundamentals of conventional (neoclassical) microeconomics. Before we delve into scale, however, let us briefly consider allocation and distribution from the perspective of ecological economics.

Ecological economists acknowledge that the market—that ubiquitous place where goods and services are exchanged—is reasonably efficient at allocating resources. The market is especially efficient when property rights are easily established and readily enforced. This is *not* the same as saying prices are a good indicator of absolute or long-run scarcity. For example, even if the market price of petroleum is far too low for the sake of the grandkids, allowing us to pull the carpet from under their future, the market will do a reasonably good job of allocating petroleum among today's power plants, airlines and trucking companies. For example, there won't be a huge surplus of petroleum at the power plant if the trucker down the road is desperate for gas. The fact that the invisible hand can handle this allocation problem is good indeed. Today's consumers will not only have electricity from the power plant, but goods hauled in by the trucker.

Adam Smith described this process in detail but also noted several problems, including monopolies and misinformed consumers. Such problems prevent the market from performing properly. Few have argued that point, and economists of all stripes talk about "market failure" and how to correct it. Nevertheless, neoclassical economists place a notorious amount of faith in the market. The invisible hand, they say, ensures that microeconomic behavior produces a desirable macroeconomic outcome. Supply and demand establish prices that send appropriate signals to producers and consumers, leading to economic activity that serves society's interests. For example, as a natural resource becomes scarcer, the price of it rises, resulting in more vigorous efforts to supply the resource.

Theoretically, this will take care of the grandkids as well as today's consumers.

Richard Norgaard, a professor at the University of California-Berkeley and past president of the ISEE, points out the fallacy inherent to the neoclassical theory of prices.[9] The theory implies that she who sells the resource knows whether the resource is scarce or not. Otherwise, how would she know where to set the price? Yet how was she supposed to know how scarce the resource was, if price was supposed to tell her? It's a catch-22.

This is an important critique, because economists often argue that natural resources are actually becoming more plentiful just because prices are declining. (Not that many prices are declining today.) The late Julian Simon (1932–1998) famously peddled such pap, spawning disciples who found Simon's argument conducive to increasing their own money supplies. After all, their "theory" feeds straight into the hands of corporations that benefit from the resulting, pro-growth mindset of consumers and policy makers. The corporate community loves these disciples of Simon, and the new darling is the Danish statistician Bjorn Lomborg. Praise has been heaped upon Lomborg by the likes of the Competitive Enterprise Institute for his book, *The Skeptical Environmentalist* (see Chapter 4).

Yet for ecologists, ecological economists and sustainability thinkers, *The Skeptical Environmentalist* is riddled with fallacies, straw men and shoddy scholarship. I agree with them, having carefully reviewed the book for the journal *Conservation Biology*,[10] and websites have been devoted to exposing Lomborg's misinformation. Yet we saw in Chapter 4 how such books can be paraded by Big Money. In the process, their popularity may eclipse their notoriety, especially among the uninitiated, the gullible or those desperately wanting to believe that all is well, after all, in the environment. George Will comes to mind.[11]

But back to Norgaard, whose observation on the fallacy of pricing theory helps explain the confusion of economics students when they encounter the subject of supply and demand in introductory "micro." First they learn that prices are determined by supply and demand. Then they learn that the quantities supplied and demanded are determined by…prices.

There happens to be no lurking inconsistency here. But there's no magic trick to dazzle us either. It's just a matter of semantics. Supply is not the same as "quantities supplied" and demand is not the same as "quantities demanded." But these semantics do open the door for shenanigans.

The supply (per se) of raw diamonds, for example, is determined primarily by how many diamonds are in the ground and the technology available for mining them. Supply clearly does influence price; diamonds are expensive partly because they are so hard to find and extract. On the other hand, the "quantity supplied" is what is brought to the market by diamond sellers. Price clearly does influence the quantity supplied; the higher the price, the higher the quantity supplied, all else equal.

So the relationships among supply, price and quantity supplied are really not so mysterious, at least not until a linguistically reckless or unscrupulous growthman wades in to muddy up the waters. The late Julian Simon has plenty of living counterparts. Robert Bradley, president of the Institute for Energy Research, believes that "natural resources originate from the mind, not from the ground, and therefore are not depletable. Thus, energy can be best understood as a bottomless pyramid of increasing substitutability and supply."[12] In other words, innovators supply the world with natural resources, including energy, from their minds. Therefore, the supply of such resources is no problem.

Clearly such a theory inculcates a healthy supply of manipulative political rhetoric, in which the word "supply" is quickly corrupted. It's a game anyone can play, so let's take a turn. Consider the supply of clean air at a party in an apartment. Smokers suck in the clean air and gradually replace it with secondary smoke. Their lungs are like pumps in an oilfield, systematically extracting the resource, replacing it with airborne sludge. As more smokers arrive, the supply of clean air noticeably dwindles, and non-smokers start to leave. Eventually even the smokers start leaving, beginning with the lighter smokers who don't like heavy smoke. So at first, more smokers means a lower supply of clean air, yet eventually—after enough smokers have polluted the place and many have left—the supply of clean air stabilizes. In fact the supply of clean air starts to increase a bit as the secondary smoke is absorbed in the curtains and carpeting, and fresh air wafts in through fissures in the walls (assuming smokers weren't crowding the hallways outside). Next, we conveniently overlook the fact that it took a major reduction of clean air to make all this happen; too complicated to consider all that. So in a squirrelly sort of way we can now say that more smoking (that is, extraction of clean air) led to *increasing* supplies of clean air, and indoor air pollution due to smoking is a self-correcting problem. If we generalize a bit, moving out of the confines of this particular party, we can say that the key to less smoking in society is more smoking!

This ludicrous example mirrors the claim that the invisible hand of the market will "fix" any resource shortages that might arise. It's smoke-and-mirrors.

We've all been downwind of cigarette smoke. Certainly we have the right to poke a little fun, especially at the "Seven Dwarves," the CEOs of America's largest tobacco companies, who perjured themselves before a US House of Representatives Subcommittee: "I believe that nicotine is not addictive."[13] The resulting news broadcast was unforgettable to many Americans, who learned a lot about Big Money that day. We fully expected the Seven Dwarves to announce, as an encore, the Tooth Fairy's engagement to Santa Claus.

So Americans know quite well how Big Money pollutes the truth. Can we expect the mother of all money-making theories, neoclassical growth theory—along with all its crazy correlates—to come to us on wings of truth? Sure, sure, higher prices stemming from lowered supplies actually "increase" supplies because they provide an incentive to "supply" even more. And more smoke makes the air "cleaner" by providing an incentive for smokers to increase the supply of clean air. More traffic increases the supply of open road.

More noise actually leads to a greater supply of quietness. Less of a good thing leads to more of it! More of a bad thing leads to less of it! Or, if you prefer, less of a good thing leads to less of a bad thing, and more of a bad thing leads to more of a good thing!

So if the Competitive Enterprise Institute, neoclassical economists and growthmen at large want to claim that oil supplies, for example, are actually *increasing*, not decreasing, as evidenced by the occasional downturn in price, let them play with the word "supply" like the Seven Dwarves play with "addictive." Let them use "supply" to mean more, less, a harmless mess, anybody's guess…whatever. But may the rest of us not be dolts. Supply is how much there is, and as you use more, less remains.

Meanwhile, expecting the market to maintain our supplies is like expecting the political arena to maintain our ethics, the library to maintain our ideas or the sewage plant to maintain our intestinal tracts. Each of these pairings represents a relationship between two variables, but in no case is the relationship straightforward or dependable, much less positively reinforcing. Thus it is with market prices and supplies. The bottom line is that markets are all about the consumption of resources. No matter how efficiently they allocate resources today, *bigger* markets mean *more* consumption and *less* resources tomorrow.

Now we turn to the distribution of wealth. Many neoclassical economists view the distribution of wealth as a final stage or special case of allocation and therefore "covered" by the market. Others think of distribution as a matter for politics, ethics or religion and not even within the purview of economics. Ecological economists, on the other hand, emphasize that an equitable distribution of wealth is necessary for the long-term economic security of rich and poor alike, and is therefore a central issue for economic study and policy. In fact, distribution of wealth generally takes a higher priority than allocation in ecological economics.

Ecological economists also emphasize the distinction between allocation and distribution. Just because a consumer purchases something when he thinks the transaction will benefit him doesn't make the market equitable at distributing wealth. In fact, the market has little to do with the distribution of wealth. Wealth (in the economic sense) is the *means* to purchase and affects the *amount* that may be purchased. Wealth may be legitimately worked for and even invested in, so that labor markets and stock markets are conduits for wealth, but large portions of wealth are distributed via charity, inheritance, marriage, luck and shades of crime ranging from shoplifting to Enron. A whole subfield of ecological economics has sprung up around the distribution of wealth as distinct from the allocation of resources.

In considering the distribution of wealth, a good starting point is human behavior. Remember from Chapter 2 the neoclassical notion of self-interested, utility-maximizing *Homo economicus*, whereby utility is expressed in terms of consumption? This materialistic model of mankind is roundly condemned by assorted critics, often with Marxist or Georgist leanings, and more often with common sense. Ecological economics offers an additional, unique and original critique in which humans are viewed as having evolved in a variety of ecosystems, each of which posed unique constraints on economic behavior and resulted in unique cultural norms. As such, humans are subject to diverse, complicated, and even mysterious motives not satisfied by simply maximizing their consumption of goods and services. This way of thinking is called "evolutionary economics" in some circles.

Don't worry, this book is not about to turn into a Luddite manifesto for turning back the clock to caveman days. But it's worth thinking about human nature—the deeply rooted, promising aspects of human nature with economic growth at the crossroads. We know that people the world over have cultural, tribal roots and urges, exposed most obviously in outdoor activities such as hunting, fishing and camping. Is there something deeper? Surely there is, especially traits, behaviors and attitudes that would have contributed to individual and tribal survival. We should at least attempt to identify some of the ways human evolution has affected our economic behavior today, rather than settling for a model that makes us look like pigs at a trough.

Early human societies, or tribes, involved kinship, a common language, a common faith, some property in common, equity among members (especially within gender), and an economy adapted to and dependent upon a particular ecosystem. Long-lasting tribal societies consisted of individuals who valued their tribal identities, including their ancestors and descendents. In other words, they were concerned with the distribution of wealth not only among the living but unto future generations. Far from maximizing consumption, they monitored their use of resources and consciously conserved these resources for future generations. Of course, not all tribes can be characterized this way, but in many parts of the world tribal institutions evolved to ensure conservation.[14] Such institutions included totems that identified clan members with non-human species, dances that reinforced appreciation of natural resources, land-resting practices and, in almost all tribal cultures, redistributions of wealth ranging in scope and duration from the Chinook potlatch to the Mosaic Year of Jubilee. Tribal cultures that failed to develop the appropriate traits and institutions were not sustainable; they simply didn't survive.

The point is that both conservation of resources and redistribution of wealth are essential for sustainability—ecologically, economically and ethically. As with the wealthy, there are needy people in all societies as a matter of luck, skill, age, health, inheritance and other factors often beyond their control. The needy perish without some help from society, or else turn to *anti*-social means of acquiring necessities. In a tribal hunting culture, the needy would have resorted to indiscriminate harvesting practices that endangered future generations of wildlife and plants, such as the killing of pregnant does. In feudal times, the needy often hid along forest paths, begging, poaching, sometimes robbing. Today's needy tend to congregate, nameless, in big cities where food and shelter are more readily obtained. If there is no assistance, whether it be some form of workfare or pure charity, eventually violence ensues. That's just common sense, and only an intelligent and fair approach to distributing wealth is stable and sustainable.

So in the long evolution of tribal cultures, institutions for maintaining equity and ceremonies for redistributing wealth were selected for because they were sustainable. That doesn't mean tribal leaders said, "Let's select these institutions because they are sustainable." It means that tribes which developed those institutions lasted the test of time, while others didn't. It was natural selection operating at the cultural level.

It should be comforting and encouraging to know that sustainable economies are not an unprecedented condition for *Homo sapiens*, especially when we consider that we all have tribal ancestry if we search deeply enough into the past. Perhaps it is in us yet to limit consumption for the sake of society, present and future, instead of attempting (in abject futility) to satiate unlimited wants. Perhaps posterity, the "seventh generation" in more tribal terms, will yet recapture our attention long enough to put the likes of gas-hogging Escalades and McMansions in a new light, a light not nearly so positive as it apparently is today.

Meanwhile, we face a troubling question: "If we were all tribal, and natural selection was for sustainable tribal institutions, whatever happened!?" The answer seems straightforward enough. During the Neolithic Period, or the New Stone Age, beginning in the Middle East around 7,500 BC, tribes learned to grow their own food. Agriculture spread shortly thereafter to parts of Africa, India, China and Europe, while Native American tribes developed their own agricultural techniques. Agriculture and the domestication of animals allowed a degree of separation from nature and independence from the wild animals and plants so important to tribal identity. It wasn't long before agricultural surplus freed the hands for the division of labor and the development of numerous technologies, occupations and cultural activities such as politics and religion. A sedentary lifestyle supported by agriculture was also conducive to larger families and higher population densities.

Friction among neighboring peoples, often for resources, resulted in the development of organized warfare. Some tribes became oriented more toward raiding than hunting, gathering or farming. Post-tribal societies similarly produced a warrior class and, eventually, national armies and navies. So the world went through its stages of empires, feudalism and monarchies; a dark age here, a renaissance there, periodically punctuated by religious crusades and revolutions. By sword or plowshare, tribal societies were replaced, one by one, often unto several post-tribal stages. For example, Polans, Silesians, and other Polish tribes were subjected to Viking invasions from the north and Mongolian invasions from the east long before there was a state of Poland, which then was invaded by Turks from the south, Germans from the west, and Russians from the east.[15] Yet the Poles retained their homeland, helped early on by sustainable tribal cultures, rooted to the land, with lasting traditions of loyalty and cooperation gradually melding with Roman Catholic ceremony.

Hebrew tribes had an even longer, more intense history of persecution. They retained their faith but lost their homelands and then their right to own land. Eventually, with no lands to co-evolve with, or to farm, hunt on, or gather from, Jewish society naturally became oriented toward commerce. Lending, especially, required almost no land. Christians were not allowed to do it, so Jews occupied, expanded and at times perfected this unique niche in the financial history of the world.[16] Money-lending (a forced occupation) may not jump out as an icon of sustainability, but other tribal traditions do, such as the labor-and-land-resting Sabbath day, the land-and-labor-resting *shmita* (the Sabbath year), and the leveling of wealth known as Jubilee. These traditions were so sustainable—protecting land and spirit as one—that they lasted centuries after the lands themselves were out of reach.

All five of the world's major religions—Hinduism, Buddhism, Islam, Judaism, Christianity—have sustainable traditions, at least ideally. Hindus revere nature and eschew a materialistic lifestyle. Buddhists follow the middle path, a perfect metaphor for the balance of nature, with humans taking their share while leaving the rest for the other species of the world. Moslems establish "hima," or nature reserves, to balance their needs with the needs of plants and animals. Jews rest the land and participate in Jubilee. Christians are stewards of nature in the mold of St. Francis of Assisi. An argument could be made that protecting the environment, out of respect for nature and concern for future generations, is the most unifying theme of the major religions. Who among the bona fide faithful would deny its importance?

I'll never forget the day I was asked to give a talk on steady state economics to religious leaders in the Washington, DC, area. During the discussion that followed, a distinguished and pensive Unitarian minister finally revealed his thoughts by saying, "You know, the steady state economy—that's the kingdom of God." He elaborated to some extent on the theological basis for this statement, yet it is easy to sense how perpetual economic growth doesn't mesh with the ideals of *any* major religion. Neither does perpetual recession. That leaves the steady state economy as the theologically enlightened alternative.

Perhaps the environmentally ideal aspects of mainstream religion stem primarily from the earthy spiritualism, common sense and dignity of sustainable tribal traditions. But ideals are rarely achieved, and not all tribal traditions lasted beyond the tribes themselves. Many Native American (North and South), Australian and African tribes were obliterated by imperialist European nations, who turned out not to represent ideal versions of Christianity, Islam or other religions. Some of the tribes remain in name, at least, within and among modern nation-states, but only the deepest Amazonian rainforest or the driest Australian outback still have tribal economies rooted intimately in their ecosystems.

Meanwhile an explosive convergence of science and technology, all in the midst of an intellectual "Enlightenment," led to the Industrial Revolution of 18th-century Europe. In the evolutionary perspective of ecological economics, the very phrase "Industrial Revolution" is telling. A *revolution* is something that, by definition, has become unhitched from *evolution*. In a revolution, the pace and magnitude of change are pronounced. Industrialization happened in a flash compared to the long sweep of prehistory, and suddenly most of the world participates in globalizing, mass-marketing, manufacturing and even "service" economies. Many of us have lost our connection to the natural world and wouldn't know a grouse from a grebe. We have lost the tribal institutions that kept us in touch with the natural resources the grandkids will depend upon. The challenge now is to develop counterparts to tribal totems, ceremonies, land-resting rules and even distributional schemes that will work in today's political economies.

Some such counterparts limp along in disguise already. In the United States, for example, the bald eagle has been our nation's symbol since 1782. The identification of our populace with this majestic bird helps to explain the strong protections afforded to the eagle, dating back to the Bald Eagle Protection Act of 1940. It's a solid perch in the mostly unsustainable tree of American policy.

Certain seasonal events evoke a touch of tribal awareness, too. Thanksgiving is probably the closest thing in American culture to a tribally rooted celebration in which we ponder and appreciate the plenitude of the land. It is no coincidence that, of all the federal holidays, this one brings us closest to our Native American hosts. We are thankful to the Native Americans for helping those early, vulnerable colonists. Alongside the Native Americans, we are also thankful to God or Mother Nature for the fruits of the land. It's true that Thanksgiving has become a lot like an American Christmas—more about celebration than appreciation. The malls are open till midnight and a lot of Americans spend the day shopping for Christmas gifts. The sheer mass of this operation has become unsustainable, but at least an element of wealth redistribution lives on in the act of gift giving.

So now in the 21st century we must stand before the mirror and ask: Which of the following ladies or gentlemen will materialize? A long-evolved, tribally rooted, *Homo ecologicus*, or the self-interested, utility-maximizing, globe-trotting *Homo economicus*? Which, we might ask, would also deserve the title of *Homo sapiens*? Clearly we need *sapience* in a full-world economy.

Based on the above—our mix of tribal origins and industrial economies—I'd like to think we are *Homo ecologicus*, variety *economicus*. We are predisposed, while remaining efficient and adequately self-interested, to distribute wealth in a more sustainable fashion than the raw-boned, sociopathic *Homo economicus*. We've got just enough sapience to be vigilant, to maintain or restore our ecological and ethical fitness, to keep in mind that the unfit go extinct, with or without piles of money.

For purposes of ethical fitness, the distribution of wealth is our primary concern. For purposes of ecological fitness, the bigger issue is scale. The market may do a reasonably good job at allocating resources among competing ends, and is involved to some extent in distributing wealth, but it does nothing whatsoever to prevent the over-allocation of the entire collection of resources or even of a particular resource. Neither does neoclassical economics. In neoclassical economics, it's unclear if land is even a factor of production (Chapter 4), technological progress perpetually increases economic capacity (Chapter 8) and population growth is required not only for long-term GDP growth, but for long term per capita GDP growth (Chapter 5)!

As with the distribution of wealth, there are tribal antecedents that enlighten our understanding of the scale issue. An oft-cited example is the Rapa Nui of Easter Island, who developed a culture obsessed with the conspicuous display of stone figures called *moai*. Moai often weighed more than 20 tons, and the desire to move them about the island resulted in a technology of transportation in which copious quantities of coconut palms were used as rollers.[17] Competition among islanders to display more and bigger moai took precedence over developing institutions for monitoring and conserving the palm, which also happened to be a crucial source of food and fiber. The Rapa Nui neglected other natural resources too, but the coconut palm was the cornerstone of their economic life. Once their island was denuded of coconut palm, disaster ensued.

The economy had become far too big for the remaining resources to support. Cultural decay set in quickly, to the point of cannibalism, until the economy was adjusted to an ecologically supportable level by a not-so-invisible hand of nature.

Anthropology is not an exact science, but it appears that prior to the Industrial Revolution many economies on continents outside Europe had achieved relatively steady states (as in steady state economies), even while others were heading down paths toward Easter Island-like outcomes. For example, several tribes in North America had developed hunting economies in balance with the bison herds roaming the Great Plains, especially before the Spaniards introduced horses.[18] Among these noteworthy tribes were the Arapaho, Cheyenne and Comanche. (The famous Sioux tribes came later to the plains in response to European colonization, moving in from the east and adapting quickly.)

Meanwhile, a more sedentary, Anasazi culture was waxing and waning in the Southwest. In Chaco Canyon (New Mexico), then at Mesa Verde (Colorado), Anasazi economies boomed and then collapsed in the 13th and 14th centuries. Scholars think the demise of these economies was largely a result of over-population and resulting resource shortages, especially of water. Not long afterward a similar fate befell the primary tribal occupants of what is now Arizona, although their wholesale disappearance from the archaeological record is a bit more mysterious. Indeed, the Pima Indians called them the "Hohokam," or "vanished ones." In the case of the Anasazi, pockets survived here and there, evolving culturally into the more sustainable Pueblo tribes of today.

Clearly, there was a natural selection for sustainable tribal cultures in North America, and a similar process was underway for millennia over large swathes of the planet that had avoided the Neolithic transformation and its discontents (the proverbial Vandals, Visigoths and Vikings). Rather suddenly and very tragically, in an early episode of "globalization," the whole process was interrupted by a lethal combination of European "guns, germs, and steel."[19] Many tribes disappeared, and those that survived lost a great deal of cultural integrity, including in many cases the institutions that had made them sustainable.

We should avoid, of course, a polyannaish perspective on prehistoric life. Even among the tribes that appear sustainable in hindsight, survival wasn't fun and games, especially for women saddled with heavy domestic workloads. Nor was peace a long-lasting condition, evidently, in regions of tribal interaction. Natural hazards were ever-present and, as many economists have emphasized, the average lifespan was much shorter than today's.

But we should also avoid the presumptions of economists who are polyannaish in their perspective of *current* affairs. Yes, of course life spans are longer today, but there is ample evidence that many a tribal life was lived in magnificent, vigorous health, especially in the hunting cultures of North America and Africa.[20] We cannot know how much vigor was lost to humankind when tribal blood stopped coursing through its veins, or how much "disutility" we experience as a result of pollution, noise and the myriad other stresses of a full-world economy. It would be ludicrous for us to claim the slightest knowledge of comparative health, happiness or general welfare. It is every bit as ludicrous for economists, à la the late Julian Simon, to conjure up such supposed knowledge.

This brings us back to neoclassical economics, which envisions the economy as a circular flow of money. Money flows from households to firms and back again in circular fashion. The circular flow of money is taught in introductory business courses but is roundly ridiculed by ecologists for its failure to reflect more than a measly amount of reality. It certainly makes tribal life look sophisticated.

The ecological critique of neoclassical growth theory begins by noting that the circular flow diagram omits a little detail called the ecosystem. The economy, as Daly pointed out, is but one subsystem functioning within the ecosystem at large. The problem with the circular flow model in neoclassical textbooks is that it fails to even mention the larger system—the ecosystem—within which the money flows. We could even argue that it is the ecosystem *from* which the money flows (and we will, in the next chapter).

It is true that many economics texts build upon the simplistic circular flow with additional factors and agents. For example, one common diagram incorporates the government as a major money handler, taking taxes from firms and households and doling out salaries to bureaucrats, among other expenditures. Another diagram will show not only the flow of money but also the broad categories of how the money is spent. For example, the diagram will show how firms procure the factors of production from households, while households procure finished goods and services from firms. The factors of production are sometimes referred to as capital, labor and "raw materials." The phrase "raw materials"—natural resources coming from the land—is about as close as the neoclassical model comes to identifying the ecosystem as relevant to the economy. In a full-world economy, this is not nearly close enough. It's akin to identifying the engine as merely "relevant" to the automobile.

The neoclassical economist might say, "Of course the economy exists within the ecosystem; it goes without saying." The problem with this excuse is that, as we saw in the last two chapters, neoclassical economics does indeed overlook and minimize the relevance of land in the production function. The landless production function amounts to the same as overlooking the ecosystem's role in the circular flow of money. This oversight would not have been so harmful during the classical era when the human economy was like a drop in the ecological bucket, but with bottled water, global warming and a burgeoning list of endangered species, isn't the oversight radically and recklessly unacceptable? In a full world, it behooves our economists, students and citizens to *emphasize* rather than trivialize the ecosystem as the foundation, matrix and backdrop for all economies.

Just as a very basic neoclassical textbook may include the simplest circular flow diagram consisting entirely of firms and households and the money circulating among them, a very basic ecological economics textbook may include a diagram with the human economy embedded in a very simple ecosystem. Picture, for example, a brown sphere labeled "human economy" within a green sphere labeled "global ecosystem." As the brown sphere grows, green space shrinks.

However, even the most basic ecological economics textbook will include one more extremely important component: the sun.

The sun is the primary source of the energy required to fuel all economies, including the "economy of nature" and its human counterpart. Millions of years ago it provided the energy for the photosynthesis of plants that eventually decayed their way into becoming today's fossil fuels. Photosynthesis continues today, providing us with biomass fuels such as firewood. The sun warms the Earth, too, creating thermal currents and generating wind energy. In the process of evaporation, the sun "picks up" the water from the seas and drops it upstream of dams, thus producing hydroelectric energy. The sun also meets our energy needs more directly via solar panels, and sometimes even more directly, as with greenhouses.

The only other significant sources of constantly flowing energy are the moon and the Earth. The moon generates tidal energy with its gravitational pull, while radioactive decay (primarily) continues to generate heat energy at the Earth's core. In a sense, we have the sun, moon and Earth feeding us energy. As the sustainability thinker David Holmgren pointed out, this is a curious fact when considered in the context of our tribal roots. It turns out that "Mother Earth, Father Sky, Sister Moon," are apt metaphors for the nurturing support we receive from the natural world. Yet there is no need to get "New Agey" about it. The religious call of "caring for creation" is probably a more relevant development in the spiritual world for saving posterity from an environmental and economic train-wreck.[21]

On the other hand, there *is* something well worth noting about the relationship of New Age philosophy to neoclassical economic growth theory. While classical philosophers and classical economists recognized limits to economic growth, the current theory of perpetual growth touted by neoclassical economists, corporations and politicians finds its spiritual counterpart in the New Age movement. The irony seems outlandish, given the tag "conservative"

FIGURE 6.2. Mountaintop removal for coal encroaches on one of the few remaining homes in what was the town of Mud, in Lincoln County, West Virginia. Credits: Vivian Stockman and Ohio Valley Environmental Coalition (ohvec.org)

attached to the most adamant pro-growthers today, but judge for yourself: New Age spiritualism is a unique combination of technological optimism and a concept we might summarize as "mind over matter," whereby "natural resources originate from the mind, not from the ground." New Agers advocate extensive genetic engineering, "astral traveling" and wispy notions of energy transformation that are unabashedly referred to as "magic" or even "alchemy." The New Age movement constitutes a fantasizing, expansionist philosophy of human destiny in which the limits imposed by nature are transcended through a change in consciousness.

Beautiful dreams are still dreams, no matter how beautiful. Unfortunately, there is no more scientific basis for the New Age vision than there is for the neoclassical theory of perpetual growth. We get a certain amount of energy from the sun's rays, the moon's pull and the Earth's core. Various useful forms of energy are derived from each of these sources. In addition to the aforementioned wind, wave and hydroelectric energy noted above, we have geothermal energy derived from the ventilation of the Earth's core. Ecological economists refer to these as "renewable" energy because they will flow from the sun, moon and Earth for a very long time.

There are still other, non-renewable sources, and two are significant: fossil fuels and uranium. We will consider these, but first note that they are moot for economic purposes in the absence of sunlight, photosynthesis and the resulting plants required for the existence of all animal and human economies. Fossil fuels and uranium may be used to supplement our energy needs, especially in the manufacturing and services sectors, but they cannot substitute for the solar energy that literally, through photosynthesis, powers the agricultural sector at the foundation of our economy. Energy income from the sun establishes an absolute upper limit to sustainable economic production, an upper limit to gross world product.

Some will argue that we can eventually replace photosynthesis with another process of food production, a process not requiring sunlight but, perhaps, only heat, so that nuclear power may be used instead. Such hog-wild fantasia makes even an ultra-liberal New-Age charlatan look like Charleton Heston (the late, ultra-conservative president of the National Rifle Association.) There will be technological developments that increase agricultural efficiency, and probably significant ones yet, but we should not allow our society to be seduced into complacency by the lunatic fringe of technological optimism.

In addition to the sources of energy, we need to understand something of the nature of energy. For this purpose we turn to the branch of physics known as thermodynamics. We encountered the second law of thermodynamics (entropy law) earlier in the chapter; we need only consider the basics a little further to grasp what the laws of thermodynamics mean to economic growth.

Thermodynamics is a branch of physics dealing with the properties and behavior of energy, especially the movement (dynamics) of heat (thermal energy).[22] The first two laws of thermodynamics are the important ones for our purposes. The first, phrased in popular terms, is that energy is neither created nor destroyed. Energy doesn't disappear and the universe has a fixed amount of it.

Energy can, however, be transformed or converted in numerous ways, many of which are relevant to economic growth. For example, we use wind turbines to convert wind energy to electrical energy. We use furnaces to convert the chemical energy of coal into thermal energy. We use bongo drums to convert the kinetic energy of a moving drumstick into a form of wave energy called "sound."

The energy transformation process ecological economists emphasize more than neoclassical economists is the process of photosynthesis, by which plants convert electromagnetic energy (light) into chemical energy with the help of a little water and soil. This is the most widespread energy transformation on Earth and supports virtually all life. Economic growth interferes with photosynthesis because it tends to replace plants with pavement (or other less-than-natural features). Not every economic activity precludes photosynthesis, but only the agricultural, silvicultural and pastoral sectors incorporate substantial amounts of photosynthesis directly in the production process. Even in many of these cases, activities like poor ranching practices in arid regions result in a negative net effect on photosynthesis.

There is another sort of energy transformation almost as profound as photosynthesis. It was identified by Albert Einstein: "It followed from the special theory of relativity that mass and energy are both but different manifestations of the same thing—a somewhat unfamiliar conception for the average mind. Furthermore, the equation E is equal to mc^2, in which energy is put equal to mass, multiplied by the square of the velocity of light, showed that very small amounts of mass may be converted into a very large amount of energy and vice versa."[23]

"Very large," indeed. For example, there are approximately 30 grams (slightly more than an ounce) of hydrogen atoms in a kilogram (2.2 pounds) of water. Einstein's formula tells us that converting those 30 grams of hydrogen would yield 2,700,000,000,000,000 joules of energy. This is the amount of energy emitted in the combustion of 270,000 gallons of gasoline!

Because energy and mass are "different manifestations of the same thing," we can restate the first law of thermodynamics: "Neither matter nor energy are created or destroyed," although matter may be *transformed* into energy. Apparently energy may be transformed into mass, too, such as when a high-energy photon passes near an atomic nucleus and is converted into an electron and a positron.

In any event, the first law of thermodynamics puts a cap on the global economy. The economy cannot be larger than what is made possible by the available matter and energy. At first glance this may seem like a highly theoretical point, yet it is an extremely important point, because it refutes the claim that there is no limit to economic growth. The only argument left standing that even resembles the no-limit claim is, "There may be a limit to economic growth, but it is so far off that we need not consider it for purposes of policy and management." Hopefully, Chapter 1 sufficed to show that the time is now to get serious about the limits to growth. If not, the rest of Part 3 should do it.

There is but one other argument remotely supporting the claim of no limits, and it goes like this: "Of course there is a limit to the production and consumption of goods and services, but there is no limit to the *value* of those goods and services. Therefore, there is no limit to economic growth after all." We shall deal with this reddest of red herrings in Chapter 7. For now, a few more observations on Einstein's discovery are in order.

$E = mc^2$ opened a lot of doors, some of hope, some of horror. Unfortunately, the doors of hope are still largely theoretical, while the doors of horror swung open immediately. If we could pry open the theoretical doors of hope, we would enter a world where the awesome potential of the atom has been harnessed to do our economic bidding *and* pose little risk to our health. The doors of horror, on the other hand, were blasted open in the New Mexico desert and the hallways led to Hiroshima and Nagasaki. There is no going back, either; only vigilant effort to prevent going further.

Nuclear technology may be used in peace and war alike, but we should remember the bomb came first. It wasn't until the 1950s when peaceful purposes of nuclear fission were developed. The United States, United Kingdom, Russia, China, France, Israel, India, Pakistan and North Korea are known to have nuclear bombs. Today there are approximately 450 nuclear reactors among 30 nations.

When it comes to $E = mc^2$, the United States can't seem to pry open the doors of hope and it can't seem to guard the horrible doorknobs from newcomers such as Iran and Libya. To add to the confusion, no one knows for sure what the United States hopes to accomplish with its nuclear technology. Self-defense? Or GDP growth in an economy that is 85 percent fossil-fueled? Self-defense would help justify the American government seeking out and destroying weapons of mass destruction, even if it seems hypocritical to most of the world. Nuclear-powered GDP growth in a full-world economy, on the other hand, is actually a *threat* to national security and international stability.

I came to the subject of energy availability in the 1990s during my PhD research. While conducting a policy analysis of the Endangered Species Act, I was analyzing the causes of species endangerment in the United States, which turned out to be a *Who's Who* of the American economy.[24] It struck me that the constant search for more energy to fuel more economic growth would simply lead to more endangered species and less biodiversity. As I suggested in *Shoveling Fuel for a Runaway Train*, when you're riding a runaway train you'd be better off running *out* of fuel, not finding a more plentiful source.

But then, some will say, we could have more powerful brakes, or could more quickly straighten the tracks ahead, if only we had more fuel to power the brakes or fix the tracks. This is akin to saying that, if only the obese had a more plentiful food supply, they could devote the extra calories to studying methods of dieting. Surely all that extra food could be used to lose weight! Do you believe it? Likewise, more energy for economic growth wouldn't be devoted to applying the brakes. To put it in less metaphorical terms, energy for economic growth, *by definition*, is used for increasing the production and consumption of goods and services in the aggregate.

Howard T. Odum, known as "H.T.," was a brilliant systems ecologist who passed away in 2002. He gradually focused on the energetic limits to economic growth as his career at the University of Florida progressed. His work on this topic culminated in *A Prosperous Way Down*, published in 2001. Despite editorial help from Charlie Hall, Odum's one-time star pupil and a tell-it-like-it-is professor at the State University of New York (Syracuse), *Prosperous Way Down* is esoteric and remains somewhat obscure even among ecological economists. Odum builds his theory around a concept he calls "emergy," which opens the linguistic door to jargon such as "emcalories," "emjoules," and even "emdollars." The basic concept is quite simple, however. Emergy is defined as the energy "that has to be used up directly and indirectly to make a product or service."[25] In other words, emergy is the sum of *all* energy embodied in a good or service. It is sometimes referred to as "energy memory."

A wooden table's emergy, for example, is equal not only to the watts of electricity that ran the table saw and lathe used in forming the wooden parts of the table, but also the solar power required to grow the tree that produced the lumber. Plus the solar power required by the ancient life that was eventually fossilized and became fuel for the chainsaw that cut down the tree and the electric plant that ran the sawmill and the shop tools. And the solar power required for growing the amount of food that gave the logger, miller and furniture-maker the calories to do the work required in the production of the table, and so on. Screws, drills and the associated miners who extracted the metals for screws and drills would all be accounted for in a thorough calculation of the wooden table's emergy.

As with all goods and services, ultimately it is solar energy that accounts for virtually all the energy that went into the production of the wooden table. Therefore, Odum invented the term "solar emcalories" as the common currency of embodied energy. While the sun's supply of energy may seem practically limitless to the neoclassical economist, Odum's emergy concept helps to illuminate just how energy-intensive, and limited, today's industrial economy is. We can't keep pumping out higher quantities of goods such as tables, Hummers and Metrodomes, or services such as massages, love cruises and Super Bowls using only our annual allowance of solar energy. We have to go to the well—the oil well—again and again, deeper and deeper, burning up the solar energy that drenched the earth those millions of years ago, burning up emergy. What happens when the well runs down, way down? This is precisely what Odum's "prosperous way down" addresses.

In Chapter 1 we briefly considered the "Olduvai theory" of energy production, the scary scene in which per capita energy production plummets after teetering at the edge of a steep gorge.[26] Odum held out hope that the social and economic adjustments to a world with dwindling oil supplies could occur gradually and gracefully enough to be, in some holistic sense, "prosperous." This "prosperous way down" would entail a gradual return of self-sufficiency and resourcefulness to the American lifestyle, with similar adjustments required in Europe, Japan and (by now) much of China, plus all of the motor-driven megalopolises of the world. For example, instead of mass markets of groceries shipped from afar, people would tend little gardens and establish little trading cooperatives.

As I write from the midst of the Washington, DC, metropolitan area, where millions of people live in apartments, townhouses and condos, and knowing the Atlantic seaboard is increasingly covered by such metropoli, I have serious doubts about Odum's hopeful scenario. What we *can* be certain of is this: assuming a prosperous way down is even possible, it is not going to happen as long as nations are hell-bent on economic growth. Hell-bent nations take hand-carts to hell, not prosperous ways down. Virtually by definition, the prosperous way down will require reduced production and consumption of goods and services: less trucking, less packaging, less marketing, etc. In Odum's terms, this means less emcalories per foodstuff, less emcalories per wooden table, and certainly less emcalories spent on Hummers and NASCAR. Lest that word "spent" go unnoticed, we are talking about less GDP.

Odum went so far as to propose a new monetary currency to assist in the ironically named "prosperous" way down: the "emdollar." The amount of emdollars paid for a good (or a service) would reflect the amount of solar energy embodied. For example, consider the amount of money paid for two tables, each identical in materials, appearance and utility. The first table is produced using chainsaws, trucks and electric lathes. The second table is produced using handsaws, horses and carving tools. In today's American dollars, the first table costs less because with fossil fuels not yet burned up by Escalades and NASCAR, chainsaws and trucks are cheap to run. Also, much less labor is required to use such machinery than to use hand saws, horses and carving tools. We may think of this labor as being subsidized by cheap gas.

In emdollars, on the other hand, the hand-sawn, horse-drawn, carved-leg table would cost less. There may be more labor required to build this table, but the calories burned by the sawyer and horse-driver and leg-carver are trumped by the enormous amount of solar energy embodied in the fossil fuels that run the chainsaw, truck and lathe.

Economists should immediately recognize Odum's proposal as an attempt to advance an "energy theory of value." Philosophers will point out that such a theory proposes that goods and services have intrinsic, inherent values. Historians will add that ever since Aristotle intrinsic value has been distinguished from "value in exchange," or the worth of a commodity in terms of its capacity to be exchanged for other commodities.

Meanwhile value in exchange is expressed as "price." What determines price became a major topic of debate among the classical economists, as we saw in Chapter 3. Adam Smith thought the major determinant of price was utility, Ricardo thought it was labor and Marx thought it was the profit motive of the capitalist. Finally the neoclassical economists, led by Alfred Marshall, developed the theory that prices are determined in a free market by supply and demand at the margin. And for us as consumers, price is an everyday practical concern as we manage our income and budget.

Nevertheless, intrinsic properties do have a major influence on price. Gold, for example, is highly priced not only because of the high demand for it, but because it is rare. The supply is low and the effective supply is much lower yet. If all the gold on Earth consisted of large nuggets sitting on the seashore, its effective supply would be much higher than if the gold were far below the Earth's surface. It would cost much less, too, because it would take *much less energy* to extract. How much do you think an ounce of gold would cost if we had to mine a mile into the Earth to find it? Why would it cost so much? Largely because of the *energy* it would take to extract it.

The neoclassical scissors of supply and demand don't quite cut it. It takes energy for the invisible hand to do the cutting. The more energy it takes, the less slicing will be done. The invisible hand wields its scissors along the paths of least resistance, but lots of supplying and demanding takes lots of energy.

In more technical terms, energy requirements are inversely related to supply. The more energy it takes to extract or otherwise produce goods or services, the lower the supply effectively becomes. If it took *no* energy to produce goods or services, presumably supplies would be limited only by the amount of *materials* required to produce the goods and services. Because all goods or services do require energy for their production, however, we see there can be no such thing as an unlimited supply of goods and services.

What are the implications of all this to Odum's work? The short answer is that Odum was a utopian if he thought the emdollar would be adopted as a medium of exchange in the face of free-market ideology. In a free market, energy requirements do affect prices because they influence effective supplies, but prices are also affected by demand. Emdollars would do a reasonably good job of reflecting supply but not of demand. Therefore, the emdollar would have to be foisted onto the market, past the invisible hand. It could be done if citizens really, really wanted things priced that way. But when entire states such as Arizona require their high school students to take courses in "free enterprise" rather than ecology, the emdollar won't make it off page one of Odum's book.

The long answer, on the other hand, will come out of our struggles to develop the policies required for the grandkids' security, because Odum's work provides some of the necessary conceptual groundwork. We may never adopt the emdollar, but we will need to develop other policy tools (for example, higher energy taxes in American dollars) that do help us get the prices right. Odum seemed a rather wise fellow, and perhaps getting the prices right—even if in regular American dollars—is what he intended all along.

But even getting the prices right isn't going to save the day with economic growth at the crossroads. Proper pricing is a microeconomic approach to a macroeconomic problem. We're getting there...

CHAPTER 7

Don't Sell the Farm: The Trophic Theory of Money

> *Agriculture...accounts for just 3 percent of national output. That means that there is no way to get a very large effect on the US economy.*
> — WILLIAM NORDHAUS

Today we hear all kinds of talk about the "Information Economy." We are, it is said, no longer in the industrial era. This transformation from the industrial to the information economy was one of the "megatrends" outlined by John Naisbitt in his bestseller of the same name.[1] In the United States and most of Europe, we've already broken through. We're in a bona fide information economy, and presumably that warrants an exclamation mark! In China, on the other hand, the industrial fires burn so intensely that an information economy might have to emerge from the ashes, assuming it does emerge. India is a curious case in which the information sector seemed to rise out of almost no ashes, with little industrial phase to bid adieu. Of course, most of the information services in India are provided to Americans and Europeans via phone line and Internet. Is there, then, an Indian information economy? Or is it a Euro-American information economy with an Indian supplier? Or is it "Chimerica," the Chinese-American saving-spending partnership described by Niall Ferguson,[2] subsidized by an Indian information sector? Does it matter?

Yes it does, because the information economy supposedly has tremendous implications for economic growth, nationally and globally. Basically, the argument goes like this: More and more of our economy involves transactions in which the product is simply information. Therefore, less and less resources are being used up in the course of economic activity.

It is an utterly fallacious argument, namely, the self-sufficient services fallacy.

A good example of the self-sufficient services fallacy appeared in an opinion piece by Katherine Ellison, "What if they held Christmas and nobody shopped?"[3] Ellison began with a timely reminder that the profligate consumption characterizing Christmas in the US has a heavy ecological footprint. She rightly noted that such consumption is finally being scrutinized by various organizations. She noted, "The rapid rise of anti-growth groups...suggests people are catching on to what one recent book dubs the fallacy of 'shoveling coal on a runaway train.'"

Actually, the book was *Shoveling Fuel for a Runaway Train*, and the title was not meant to dub a fallacy but to introduce a metaphor. (I know because I wrote the book.) The "runaway train" is the American economy, and "shoveling fuel" describes the effect of conspicuous consumption. When we're on a runaway train, heading for a wreck, the last thing we ought to spend precious time on is shoveling fuel!

Getting the title of my book wrong was a minor gaffe, but she was just getting started. Ellison described an interview with Herman Daly, who defended the merits of a steady state economy and then "tried to turn the tables, asking, 'What do *you* [Ellison] think the future is going to look like?'" Ellison responded, "I'm not really looking past the holidays." That's a very human, humble acknowledgement, and would probably resonate with many busy readers. If she had looked past the holidays, however, an admirable New Year's resolution would have been to learn a little steady state economics!

Instead, Ellison speculated, "In that cowardly spirit, here's my compromise. This winter, I plan to support the US service economy. I may just buy mom a massage, give my kids an hour of rope-climbing, and find a personal trainer for my husband.... I can help keep the world economy chugging without contributing to all those greenhouse-gas emissions..." With this reasoning, Ellison committed the self-sufficient services fallacy. It is probably the most common fallacy among those claiming there is no conflict between economic growth and environmental protection. The key point in debunking this fallacy is that the service sectors—including massages, rope-climbing and personal training—are part of an economy that grows as an integrated whole. This will be a common theme over the next two chapters.

To Ellison's credit, she did not use the self-sufficient services fallacy to promote economic *growth*, at least not explicitly. She used the non-committal "keep the world economy chugging" rather than "keep the world economy growing." An economy may chug at a sustainable level; indeed that's a steady state economy. Unfortunately, if we use the self-sufficient services fallacy to promote economic chugging (or anything else, for that matter), we empower others to use the fallacy to promote economic growth. The difference would be one of degree and not of principle.

For example, one could say, "I plan to support the service economy even more than Katherine Ellison. I'll buy mom *five* massages, plus all kinds of information. I can help keep the world economy *growing* without contributing to all those greenhouse gas emissions." That would be wrong, and dangerously so. Unfortunately, the self-sufficient services fallacy appears to be a seductive argument for many, many people (especially politicians). That is because equally many people have not studied ecology, in particular the concept of trophic levels.

The best way to demonstrate the concept of trophic levels is with a simple diagram (Figure 7.1). Trophism refers to the transfer of energy and nutrition from one organism to another in the process of feeding. In the economy of nature, only plants produce their own food, with the process of photosynthesis. The growth of plants is called "primary production." All animal life depends on the plant

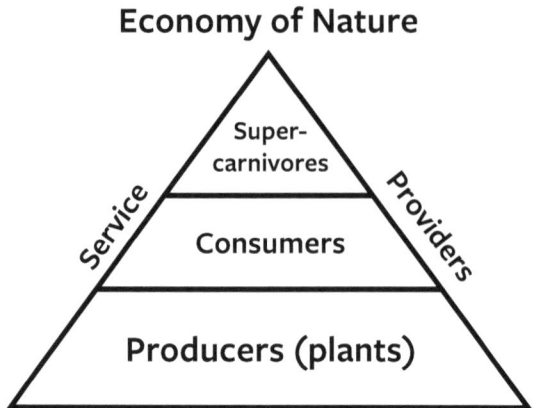

FIGURE 7.1. Trophic levels in the economy of nature. Service providers interact throughout all levels.

community for nutrition. Some animals eat plants directly, some eat animals that eat plants, and some eat animals that eat animals that eat plants. Each of these levels has a name: producers (plants), primary consumers (animals that eat plants) and secondary consumers (animals that eat those animals). Species in the highest level, such as lions, eagles, crocodiles and sharks, are sometimes called "super-carnivores."

Sometimes the precise location of a species in this system is nebulous. For example, a coyote living in one geographic area may subsist almost entirely on plant materials (grasses, tubers, berries, nuts, etc.), while a coyote living in another area may subsist primarily on small mammals and birds. It becomes even more difficult when some of the super-carnivores are considered. In many areas during the spring, grizzly bears are vegetarians. In many of the same and other areas, they specialize in harvesting salmon, which themselves are predators. So in the course of a few months, a single grizzly can go from being a primary consumer to a super-carnivore. And of course this can even happen during the course of a day, as when the bear locates a productive berry patch along the shore of an equally tempting salmon stream. However, such difficulty in categorizing species according to their trophic levels does not reduce the applicability of this concept to the human economy. In fact, it makes the two economies even more analogous, as we will see.

In the economy of nature there are also a wide variety of species that do not easily fit into any trophic level at any time. These include bacteria, worms, bumblebees, leeches... small invertebrates, for the most part. These species have odd ways of making a living. They neither produce nor consume, at least not in a predatory fashion. Some of them are parasites, but virtually all are beneficial to the economy as a whole.

A large percentage of these species (myriad bacteria, for example) make their living by decomposing plant and animal materials that are either too small, too spoiled or otherwise too indigestible for "regular" consumers. Were it not for them, the Earth would rapidly become a heap of organic rubble. Some of them, like bumblebees, ingest minuscule amounts of plant nectar. In the process, moving from flower to flower, they pollinate these plants, and without them many plant species would go extinct, eroding the base of producers. Some of them, like earthworms, ingest undifferentiated organic matter. In the process, they unwittingly till the soil, making it more porous for water infiltration and efficient root growth. All of these types of species, in essence, provide services to the economy as a whole. Depending precisely on how you distinguish these service providers from "true" consumers, they constitute a high proportion of species.

The human economy also consists of trophic levels (Figure 7.2). This has been recognized in some sense at least since the 1760s when Quesnay set out to demonstrate that the true producers in the human economy were farmers (Chapter 3). Farmers, in other words, comprised the producer trophic level in the human economy, although Quesnay did not put it in terms of trophic levels. Later economists disagreed, first arguing that labor applied in other (non-agricultural) activities was also productive, then arguing that capital itself was. The arguments about what truly constituted "productivity" among the likes of Adam Smith, David Ricardo, John Stuart Mill, and Karl Marx boiled down to a matter of semantics

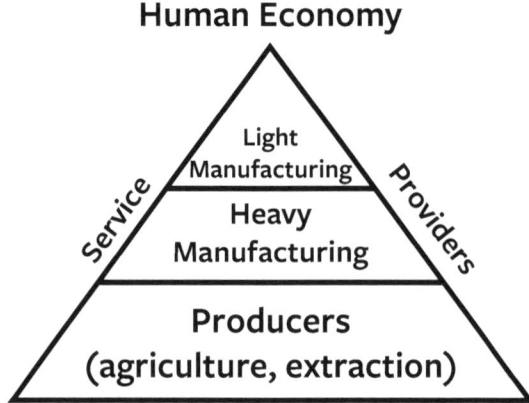

FIGURE 7.2. Trophic levels in the human economy. Service providers interact throughout all levels. Not shown is the foundation of "natural capital" that is farmed and extracted.

(albeit with ideological intent in Marx's case at least). Perhaps such argumentation could have been avoided, or at least relegated to an appropriately lesser notch of importance, if only Quesnay had gone a step further. The fact is that the farmers themselves are not quite the ultimate source of productivity either. Just as in the economy of nature, the *plant community itself* best qualifies for the title of "source." Vegetarians or not, all animal species (including *Homo sapiens*) depend on the plant community for life. This will be so unless technology is developed to create entirely synthetic foods, in which case the consumers won't quite be human.

In the human economy, most members do not make their living by literally eating what exists in the next lowest trophic level. Instead, the bottom level consists of a variety of resources that many humans harvest to make their living. In addition to plants, these resources include minerals, petroleum, fish and—today—even water (Chapter 1). Most of these resources are not even living, so it would be inappropriate to call the entire collection producers. Only the plants actually produce their own food. And some of these resources (fish, for example) exist at a higher trophic level in the economy of nature. Nevertheless, what these resources all have in common is that they comprise the foundation of materials upon which the rest of the human economy is built. In ecological economics, these materials are often called "natural capital." So the lowest trophic level of the human economy is natural capital or, in less fancy terms, land. This terminological variety has the advantage of resonating with neoclassical economists as well as ecologists, farmers and common sense.

With natural capital at the base of the human economy, the primary consumers are farmers, loggers, miners, ranchers, oilmen, fishermen and others who harvest goods directly from the land. Among these primary consumers, the farmers come closest to being true producers (à la Quesnay) because they participate closely with the process of photosynthesis.

The manufacturing trades, on the other hand, are clearly two steps removed from the foundation of the economy (natural capital), because they harvest nothing. They use the raw materials extracted by the primary consumers to manufacture goods. They range from a heavy manufacturing base (such as iron ore refining) up through the trophic pyramid to the lightest manufacturing sectors (such as computer chip manufacturing). Heavy manufacturing requires the rawest of materials, whereas much of the light manufacturing can be done with raw, refined or manufactured materials flowing from lower in the trophic structure.

In the human economy, the service sectors also defy placement in a particular trophic level. Truck drivers, bankers, waitresses, janitors, gravediggers—none produce or consume in a systematic fashion that proceeds upward from one trophic level to the next. The truck driver may deliver a load of cotton from farm to factory one day, and a load of fence posts from factory to farm the next. The banker may lend to the farmer or to the industrialist. Waitresses wait on farmers, industrialists and bankers. All contribute to GDP.

As described in Chapter 2, GDP is simply a measure of the scale of human economic activity, and it depends on how many humans are economically active and how active each one is. An analogy to GDP in the economy of nature is the amount of biomass production, biomass being the sum total of living flesh.

The growth of biomass on Earth got off to a slow and tentative start. The economy of nature apparently started with a "primeval soup" in which, by act of God or random chance, a chemical reaction involving carbon apparently produced a self-perpetuating and therefore living form.[4] Some creation theorists attribute the beginning of life to lightning, a sort of biological big bang, which they say provided the energy to catalyze this reaction. Perhaps in a series of chemical "experiments," various forms of life blinked in and out of existence for millions of years. As they say, the rest is history, albeit natural history, and today's global biomass is approximately 2 trillion tons.[5] It's not increasing ad infinitum, however. Rather, biomass and species diversity have waxed and waned for the past 540 million years, punctuated by five great episodes of extinction.[6]

It bears repeating that nature's GDP has not increased ad infinitum, nor was it ever slated to. Neither is the human economy. In fact, this is probably the right time to offer readers a sound-bite, radio-friendly refutation of perpetual economic growth. You'll win the debate with it every time. To think there is no limit to growth on a finite land mass (Earth, let's say) is precisely, mathematically equivalent to thinking that one may have a steady state economy on a perpetually *diminishing* land mass. In other words, we could gradually squish the $70 trillion global economy into one continent, then one nation, then one city…you get the picture. It's becoming an "information economy," right? So eventually we could squish it into your iPod, leaving the rest of the planet as a designated wilderness area.

Have you ever heard anything so ludicrous? Yet it's precisely, mathematically as ludicrous as thinking we could have a perpetually growing economy on Earth.

Let's look a bit more at biomass and then apply some ecological principles to the human economy. Biomass is analogous to GDP because, in nature, virtually all activity is economic, and no biomass is inactive. Unlike certain stocks of manufactured capital, such as sheetrock or fence posts, biomass can't just sit there idle. Of course there are shades of exceptions, such as a hibernating reptile or the bark of a tree, but in general the life of a nonhuman is a perpetual struggle to obtain the resources required for survival and reproduction. Ecology, therefore, is primarily about the allocation and distribution of resources in the economy of nature. Ecologists are the economists of nature.

If the "success" of the human economy is measured by its level of activity, or GDP, then presumably the success of the economy of nature may be measured by its level of activity. And that is best measured by the amount of biomass. One may wonder about the propriety of using biomass as a measure of "success." After all, what if we compared 40 billion tons of algae and bacteria with 40 billion tons of gorillas and humans? However, this question does less to negate the analogy than it does to negate the use of GDP as a measure of success in the human economy. GDP does not differentiate whether the economy is one of poor tenant farmers and a handful of wealthy landlords, or the more diverse economy we currently experience. The useful thing about GDP is that it does indeed provide a gauge of the economy's growth, regardless of whether or not the growth is a good thing.

Theoretically, a growth in biomass may come strictly from more plants, but growing biomass typically means that the ecosystem is growing as an integrated whole. Therefore, a growing economy of nature means more consumers as well as producers. Likewise, while growth in GDP could theoretically come strictly from more farmers, a growing GDP typically means that the economy is growing as an integrated whole. Therefore, economic growth means more manufacturing and services as well as more farming.

"Success" may also be measured within various subsets of the economy. We may say, for example, that an increasing proportional contribution of entertainment to GDP makes it a successful sector, just as we may say that an increasing proportional contribution of cervids (antlered mammals) to biomass makes it a successful sector. Viewed in terms of these subsets, "success" seems like a more pertinent concept than it does as applied to whole economies. After all, the entertainment sector grew because its employees successfully competed with other sectors, like the restaurant industry, for resources. Meanwhile the cervid sector grew because its species successfully competed with other sectors, like bovids (horned mammals), for resources.

Finally this leads us to the implications of trophic levels for perpetual economic growth. One of the fundamental principles of the economy of nature is that no trophic level may consist of more biomass than the underlying trophic level. In other words, the success of any one trophic level is dependent upon the success of the underlying trophic level. This follows simply from the second law of thermodynamics—the entropy law—and from the life histories of species.

The second law of thermodynamics, to put it in the simplest of terms, is that all things tend to disorder.[7] It takes energy to organize anything, whether it's a steel beam with the energy derived from coal or a cervid's antlers with the energy derived from grass. But neither of these products will last forever. They ultimately break down into a collection of substances with less order, or less embodied energy, because some of the energy is dissipated into the environment as the product breaks down. If this were not the case, the Earth would gradually be replaced by a giant collection of everything that had ever been converted from its elements. Instead, in the real world, it's "ashes to ashes, dust to dust" and along the way some piles of rust.

Conversely, the construction of these products took more energy than was finally embodied, because energy was dissipated along the way. In smelting the iron, much of the coal's energy was dissipated as heat. In growing the antler, much of the plant's energy was likewise dissipated. Thus the ecological principle that no trophic level may consist of more biomass than the one upon which it feeds.

Not only must one trophic level contain less biomass than the underlying trophic level, but there are limits to the *fraction* of an underlying trophic level's biomass that may be attained by the overlying trophic level. This too follows from the entropy law, which essentially states that nothing is perfectly efficient. These proportions are not readily ascertained; we can never expect to know precisely how much antler may be produced in proportion to how much browse is consumed, or how much iron may be smelted in proportion to ore and energy consumed. The precise proportion is purely academic. For what it's worth, though, academics who study this subject of "ecological efficiency" indicate that each trophic level contains approximately ten percent of the biomass of the next-lower trophic level. For example, in an ecosystem with 10,000 tons of producers, one may expect approximately 1,000 tons of primary consumers and 100 tons of secondary consumers. In a simplified example with only three species and three trophic levels, we might have 10,000 tons of grass, 1,000 tons of elk and 100 tons of mountain lions. Perhaps it is even possible to have 10,000 tons of grass, 2,000 tons of elk and 400 tons of mountain lions. Perhaps. But certainly we cannot have an ecosystem comprised of 10,000 tons of grass, 10,000 tons of elk and 10,000 tons of mountain lions, much less an ecosystem comprising 100 tons of grass, 1,000 tons of elk and 10,000 tons of mountain lions. There is a limit to efficiency (second law of thermodynamics) and, even more fundamentally, a limit to matter and energy (first law of thermodynamics). You can't convert 10,000 tons of grass into 10,000 tons of elk because that would entail absolute efficiency, violating the second law of thermodynamics. Likewise, you can't convert 10,000 tons of grass into 100,000 (or even 10,001) tons of elk because that would entail something from nothing, violating the first law of thermodynamics.

In the economy of nature, the life histories of animals also contribute to the "inefficiency" with which one trophic level's biomass is converted to the next. Elk, for example, expend a great deal of energy at various life stages in looking for mother, playing, escaping insects, wallowing, dispersing, fighting, courting and mating (plus, for females, raising their young). If all the bull elk's resources were devoted to maximizing the efficiency of antler growth, it would come at the expense of its other activities, including the primary advantage of growing the antlers (that is, successful courtship).

This would hardly be efficient in a holistic sense. If all the resources of a steel manufacturer were devoted to maximizing the efficiency of the smelting process, it would have none left for its other activities, including the primary reason for manufacturing the steel, selling it. Efficiency is a slippery concept, when viewed in a holistic and practical sense.

The service providers, too, are limited in proportion to the trophic levels with which they interact. Bumblebees do not live without flowering plants, unless they evolve a whole new way of living (in which case they tend to become different species, not bumblebees). Meanwhile, plants that have become dependent upon bumblebees for their pollination do not live without bumblebees, unless they evolve a different mode of pollination, including perhaps self-pollination. Similarly, chainsaw mechanics do not live without loggers, unless they evolve a new way of living (in which case they become a different economic species). And vice versa with loggers, unless they adapt to maintaining their own saws completely. This means that the amount of bumblebee biomass is dependent on the biomass of flowering plants, while the GDP contribution of chainsaw mechanics is dependent on the GDP contribution of loggers.

What all this means to the human economy is precisely the same as it means to the economy of nature: just as the capacity of the economy of nature is based on the amount of primary production, the capacity of the human economy is based on the amount of natural capital. Within this economy, the production of the manufacturing trophic level is dependent on the production of the primary consumers—the farmers, miners, loggers and such. The service providers depend on the whole system.

Is the empirical evidence consistent with this theory? Of course it is. People don't eat unless the farmer and fishermen do their jobs. That doesn't mean the GDP figures will stack up neatly in a pyramid of trophic levels. For example, a pile of two-by-fours costs more than the tree from which it was milled, and a house costs more than the two-by-fours required for its construction. On it goes through all sectors of the economy, the "value-added" product "contributing" more to GDP than the natural capital. For two centuries this added value has been attributed primarily to labor or capital. The mill worker added value to the log by milling it into a pile of two-by-fours, and the construction worker added value to the pile of two-by-fours by constructing the house. The effect is to veil or distort the trophic levels in the human economy, such that the GDP attributed to agricultural products and other natural capital is actually less than that attributed to manufactured products, and far less than that attributed to the service sectors. Therefore, we shouldn't be surprised if a single television episode of *American Idol* "contributes" more to GDP than one seasonal episode of Iowa's corn crop. The former is good for full belly laughs, while the latter is only good for filling the bellies. Right now we pay a lot more for the former.

This modern-day mismatch between trophic levels (with profound value at the bottom) and GDP figures (with big money spent at the top) has led neoclassical economists astray. There seems to be a neoclassical sucker born every minute. For example, William Nordhaus, Sterling Professor of Economics at Yale University, famously stated: "Agriculture, the part of the economy that is sensitive to climate change, accounts for just 3% of national output. That means that there is no way to get a very large effect on the US economy."[8] Herman Daly traced a succession of nearly identical errors,[9] at one point even committed by Thomas C. Schelling, a past professor of economics at Harvard, past president of the American Economic Association and 2005 Nobel laureate. In a 1997 issue of the prestigious *Foreign Affairs*, Schelling persuaded readers not to overreact to climate change by stating, "in the developed world hardly any component of the national income [GDP] is affected by climate. Agriculture is practically the only sector of the economy affected by climate, and it contributes only a small percentage—3% in the United States—of national income. If agricultural productivity were drastically reduced by climate change, the cost of living would rise by 1 or 2%, and at a time when per capita income will likely have doubled."[10]

Let's not be sidetracked by the context of climate change. It wouldn't matter if the agricultural decline was from climate change, a population explosion of woodchucks or a farmland invasion of space aliens. The salient point is that Nobel laureates with no background in ecology are talking about per capita income doubling while agricultural productivity is "drastically reduced." And not just talking over a beer at a backyard barbeque. Rather, talking in *Foreign Affairs*, giving influential policy advice. It would seem ridiculous enough to be funny, if it didn't put us in such serious trouble!

This misleading distortion—percentages of GDP increasing as we move from the most to the least essential of economic sectors—compels me to advance what I would like to coin, so to speak, the "trophic theory of money."[11]

Few economists have examined the origins of money, at least not in the sense of "origins" that is satisfactory for our purposes. Adam Smith devoted Chapter 4 of *The Wealth of Nations* to the origins and use of money, but the portion dealing with the *origins* of money, including the *preconditions of its existence*, was limited to the first two paragraphs. Keynes's biographer described how Keynes "succumbed repeatedly to his 'Babylonian madness'—an essay on the origins of money,"[12] but this was really a study in historical numismatics (the study of currency) and metrology (the science of measurement). Other great minds have likewise given short shrift to the real origins of money. Rupert Ederer attempted to summarize these accounts in *The Evolution of Money*,[13] but went on to focus on the properties and use of money. Economics texts today totally disregard the origins of money. Chapters on the "creation" of money focus on the injection of money into the economy by national banks. That's like focusing on the grocery store as the origin of milk.

Let us be perfectly clear. The real origins of money were in the agricultural surplus that freed the hands for the division of labor. This made money a meaningful concept. Adam Smith alluded to this, but didn't emphasize or clarify it, and didn't have the benefit of trophic theory to do so. Prior to agricultural surplus, no one got to focus on spinning cloth, building houses, accounting for anything, writing books, dancing with the stars or doling out legal tender. That's the trophic theory of money in a nutshell, and it's just as relevant today as it was in the early stages of human evolution. Today as then, it is only when someone else produces our food that we are free to think about clothing and shelter, much less dancing, accounting, writing books or paying for anything. Without an agricultural surplus, our hands are on the plow, not on the keyboard and certainly not on a meaningless wallet. Our feet are in the field, not on the treadle, not on the floor of the stock exchange and certainly not on the dance floor.

The trophic theory of money has much more to offer, however, than basic insight about the evolutionary origins of money. It also tells us that the real (non-inflated) money supply today is in direct proportion to the amount of agricultural surplus. When a stock market crash, a "liquidity crisis" or a fiscal impasse strikes at the heart of economic growth, we had better look deeper than derivatives peddlers, bailed-out bankers or careless Keynesians in the government. The usual suspects from the financial and fiscal sectors are problematic, all right, but these financial and fiscal crises are becoming increasingly *real* as we approach limits to economic growth. The real money supply, reflecting the production and consumption of real goods and services, can only grow so far. Forcing it to grow further results in nothing but inflation.

Neoclassical economists who discount the importance of agriculture have clearly not evolved to comprehend the implications of trophic levels. Ecological economists have, for the most part. Still, I wish to take this chapter one step further, into implications that even most ecological economists have not yet fully comprehended. The trophic theory of money implies that real (non-inflated) GDP is a reliable indicator of the amount of agricultural surplus and of the "ecological footprint" of the human economy. Not a direct measurement, but a reliable *indicator*. To establish this implication, a closer look at the ecological footprint concept is required.

The ecological footprint is a measure of our demand on the planet. It is expressed as the acreage of land (and sea) required for regenerating the resources we consume and for absorbing our pollutants. Ecological footprinting makes it possible to estimate how many planets it takes to support us with a given lifestyle. It is extremely important to bear in mind that there is but one planet—Earth—known to be conducive to the human economy.[14] At this point in history, the best available ecological footprinting research indicates that we use the equivalent of approximately 1.5 Earths to provide our resources and absorb our pollutants. In other words, it now takes the Earth one year and six months to regenerate what we use in a year. Of course some of us (such as average Europeans, Japanese and especially Americans) have a far larger ecological footprint than others (such as average Indians, Kenyans and Bhutanese). But the matter of international equity is for Part 4. Here we are focused on the relationship between GDP and agricultural surplus, and thenceforth the ecological footprint.

To establish the relationship between GDP and the ecological footprint, let us start from the lower extreme: if there were no humans on Earth, and therefore no human economy, by definition the ecological footprint would be zero. So far, so good!

Now let us consider the earliest stages of hominid evolution, when humans struggled among their fellow mammalian species for the basic habitat components of food, water and cover. Was there an ecological footprint at that point in prehistory? Some would say yes, there was a small and growing ecological footprint, while others would say that humans were just part of the economy of nature, and that an "ecological footprint" was as yet irrelevant. This is a matter of semantics and irrelevant for our purposes. We are concerned with the relationship between the money supply and the ecological footprint. Such a relationship did not exist prior to widespread agricultural surplus, when money came into being.

By the time we humans got to the point of using money, the concept of an ecological footprint was quickly becoming relevant. In fact, the earliest forms of money were themselves agricultural commodities, such as the shekel in ancient Mesopotamia, which originated as a unit (approximately 180 grains) of barley. The amount of barley produced was a function of the amount of land irrigated along the Tigris and Euphrates Rivers. The amount of Mesopotamian land irrigated or otherwise occupied and managed by humans was one of the first recorded indications that humans were vulnerable to the limits of their ecosystems. Today, when the process of human-induced desertification is discussed in scientific circles, Mesopotamia is cited as the quintessential precedent. It is no coincidence that Mesopotamia is also one of the first regions to be mentioned when discussing the history of money.

Over the course of a few thousand years, the shekel evolved into units of silver and gold. More money, then, meant more mining, which itself would clearly indicate a growing ecological footprint. More importantly, though, many of the silver and golden shekels were spent on barley and other agricultural products. Metallic shekels had value because they were accepted for the purchase of food, raw materials, clothing and other finished goods and services. The production and consumption of each of these goods and services took their bite out of Earth, and the increasing flow of shekels reflected the growing ecological footprint.

Eventually, of course, shekels were also spent on arms, ammunition and all the accouterments of colonization and national defense. In other words, the governments of empires "got money" and took over its management. Meanwhile, it is impossible to imagine a war without an ecological footprint. More shekels spent by the government on warfare, along with more shekels spent on private goods and services by individuals, continues to indicate a growing ecological footprint.

Eventually money evolved (or devolved, depending on the perspective) into paper, but the way it was used barely changed at all. Money is valuable because it is legally tendered for goods and services, private or public. The use of more money indicates an increasing volume of goods and services.

The connection of a growing money supply to a growing ecological footprint should be coming into focus by now. There are but three phenomena that might distort or delay our focus. Moving from the simplest to the most complex, these phenomena are: inflation, technological progress and "animal spirits" (a Keynesian term).

Inflation of course refers to a rise in prices. When prices increase, your money buys less. As inflation progresses, you start to lose confidence in your money. If inflation runs rampantly into a condition of "hyperinflation," your money becomes worthless. When money becomes worthless, you and fellow citizens get angry and frustrated and you panic. Social and political upheaval is sure to follow. Inflation is a monster, and economists of all ilks recognize it as such. Inflation was the economic origin of the Third Reich. Inflation is precisely what happens when a monetary authority (such as the Federal Reserve System in the United States) increases the money supply faster than the real economy can grow. Recent periods of rapid, real economic growth (such as we had in the latter decades of the 20th century) have tended to result in inflation, because the monetary authorities are too removed from the realities of economic life to understand the ecological limits to growth. Monetary authorities sometimes complain about a lack of "consumer confidence." In a full-world economy it is probably more appropriate for consumers to complain about the childish "confidence" of monetary authorities, which leads to inflation.

For our immediate purposes it is necessary to acknowledge the simple fact that inflation can cloud the tight relationship between real GDP and the ecological footprint. However, the cloud is quickly lifted when we specify that we are talking about "real GDP," or GDP adjusted for inflation. And it makes little sense to speak of "unreal" GDP, or GDP not adjusted for inflation. GDP was always intended to indicate the level of production of real (not unreal) goods and service. This level of real production is accurately reflected by monetary expenditures and income only if the monetary unit is not inflated or deflated. Allowing inflation to shroud the linkage between GDP and the ecological footprint could only happen in amateur circles, but it has to be mentioned here and now set aside.

Technological progress is another story. It doesn't take an amateur to become befuddled by the implications of technological progress. Technological progress allows the same amount of natural capital to produce a greater amount (or value) of goods and services. With technological progress, apparently, economic production may increase without a growing ecological footprint. In other words, real GDP may increase without a growing ecological footprint. Theoretically, we could reconcile the conflict between economic growth and environmental protection with technological progress. So let's just keep progressing technologically and we can continue to grow the economy, with no additional environmental impact.

If you smell a fish, you have a good nose. We will explore your olfactory savvy in Chapter 8 and digest the relevant findings. As an hors d'œuvre, let us recall that there's no such thing as a free lunch. Technological progress is not free, and its costs add up in real GDP. Ultimately, technological progress is limited by the laws of thermodynamics. (Remember, we can't produce something from nothing, and we can't get 100 percent efficiency.) This leaves us with only animal spirits shrouding the relationship between GDP and the ecological footprint.

I am taking a bit of rhetorical license here, because "animal spirits" was coined by Keynes to describe the emotions or attitudes of consumers. Here I am adapting the term to describe not only the "propensity to consume," as Keynes called it, but the propensity to *use money* in order to consume. Even in the most modern of monetary economies, the use of money is not necessarily required for consuming things we find valuable such as friendliness or compassion. It is our common sense or "animal spirits" that tell us when it is appropriate to use money for procuring satisfaction. Using the term "common sense" reflects stability in our judgment of when to use money; using the term "animal spirits" reminds us that our judgment may be altered (or may falter) at times. As with inflation and technological progress, animal spirits could shroud the relationship between GDP and the ecological footprint.

For example, if two billionaires were determined to prove that there is no relationship between GDP and the ecological footprint, they might say, "Let us now pay each other a billion dollars apiece for saying the word "ombudsman." One billionaire would say "ombudsman" and the other would pay her a billion dollars. The latter would echo "ombudsman" and be paid back the billion dollars. On and on the utterances of "ombudsman" would go until, by the end of the day, a trillion dollars had been "spent" on utterances of "ombudsman." If each billionaire claimed that the utterance of "ombudsman" was a finally produced good or service, would our national income accountants argue? They might, unless the political pressure to demonstrate GDP growth was irresistible.[15] And for those vested in perpetual economic growth theory, the temptation would be difficult to resist, politics or none. After all, a trillion dollars would have been spent—even "earned"—in one day among two people, to prove that we could "dematerialize" economic growth.

Imagine if everyone with time on their hands spent the day exchanging money for utterances of "ombudsman!" And imagine that such expenditures were added to the official calculations of GDP. That would shoot the trophic theory of money, for it would disengage the relationship between GDP and the ecological footprint.

Of course, no one spends the day uttering "ombudsman," nor does anyone spend money on such utterances. The monetary animal spirits aren't crazy enough. Nor would you pay a friend to say "hi." In fact, you wouldn't even use money to pay your husband or wife, boyfriend or girlfriend for giving you information about the weather, dinner ingredients or their state of mind. Nor would such non-paid activities have a significant ecological footprint. Such are the animal spirits—and common sense—with regard to the use of money.

In other words, "real" expenditures go toward real things—real goods and services—that have real ecological footprints. Real expenditures do not go toward non-material things with no ecological footprints.

To be more precise, expenditures might go toward non-material, unreal things, but only for short unsustainable periods of time. This may occur, for example, when unscrupulous salesmen stir up demand for unreal "assets" such as derivatives. Soon enough, however, analysts (including consumers with common sense) conclude that these assets are actually unreal. Then the markets for these unreal assets crash and, if enough suckers bought in, we find ourselves in the midst of financial crises. Sure, not much of an ecological footprint would be associated with the "increase" in GDP, but no real increase in GDP occurred to begin with. That is why the markets crashed, back down to Earth, back down to the real economy of goods and services, produced and consumed by real people with common sense.

In a sense, the use of money is a type of social contract, not only between citizens and government, but between consumers and producers. In the classical, political social contract (à la Thomas Hobbes and John Locke), citizens gave up sovereignty to a central government in order to procure social order through the rule of law. Eventually this social contract included the creation of a monetary authority, such as the Federal Reserve System in the United States. However, the monetary social contract goes beyond the relationship between citizen and government. Pursuant to the monetary social contract, not only do citizens give up their sovereignty to the monetary authorities, but consumers give up purchasing power (in the form of money) to producers, with the understanding that what is produced will benefit them (the consumers) in a real, tangible fashion.

When a political social contract is deemed violated by the citizens en masse, a revolution or anarchy ensues. When a monetary social contract is deemed violated by consumers en masse, and whole classes of "products" (such as derivatives) are found to be bogus, a financial crisis ensues. When a government is complicit in a bogus monetary social contract (for example, by investing tax revenues in derivatives), a crisis in political economy ensues. In any event, when bogus production and consumption (such as payment for the utterance of "ombudsman," or for derivatives) become widespread, the monetary social contract is violated, markets crash, inflation ensues and real GDP is brought back down to Earth, whereupon it once again reflects the ecological footprint of the human economy.

In short, because of the trophic structure of the human economy, GDP provides a reliable indicator of the ecological footprint. To some degree, the relationship between GDP and the ecological footprint can be muddied by inflation, technological progress and "animal spirits." However, inflation is easily accounted for, so that the relationship between *real* GDP and the ecological footprint may be muddied only by technological progress and animal spirits. If technological progress rained down like manna from heaven, it could disrupt the relationship between real GDP and the ecological footprint. However, technological progress does not really rain down, and that will be the principal subject of Chapter 8. Meanwhile, animal spirits are kept within a range of common sense by an invisible hand of sorts. The invisible hand won't be doling out real money for utterances of "ombudsman," but rather for real goods and services with real ecological footprints.

Finally, however, there may come a time when real GDP, measured as it is by real income and expenditure, *declines* while the ecological footprint continues to grow. This is not a distortion of the relationship between GDP and the ecological footprint caused by inflation, technological progress or animal spirits. Keep in mind that those types of distortions occur when GDP is *growing*. Rather, this new reality, with GDP declining, is reality at its sternest. It's

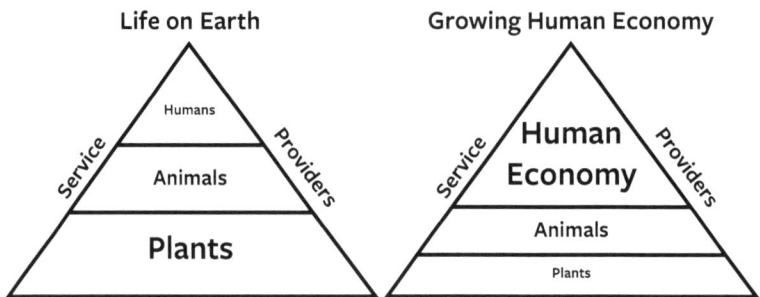

FIGURE 7.3. Trophic levels of all life on earth (*left*), making it plain to see the effect of a growing human economy—real and monetary sectors—on other species (*right*).

what happens when the human population continues to grow, placing a heavier burden on the planet, while agricultural surplus decreases, diminishing the amount of real money available for purchasing real goods and services. Is it sounding familiar enough to resonate, with economic growth at the crossroads?

We are in the midst of a full-fledged, post-industrial, 70-trillion-dollar information economy. We saw in Chapter 2 how we are pushing the agricultural limits of Earth to the breaking point. If per capita agricultural production declines far enough, the masses will be forced back to the farm and agriculture will constitute the focus of human economic activity, accounting for the lion's share of GDP, as it was in the early stages of monetary economies. If global per capita agricultural production declines to a level of mere subsistence or less, the monetary economy will virtually cease to exist, blending instead with the economy of nature where money is meaningless. All the dollars, yen or pesos in the world won't buy the last cob of corn from the farmer's field.

Yet this hypothetical example, whereby money becomes universally meaningless, should not be interpreted as a doomsayer's prediction. I for one am not predicting an ecological and monetary calamity of that magnitude, although others have done so.[16] This extreme hypothetical example would only become reality far beyond the crossroads where we currently find ourselves. Surely we won't stagger blindly straight ahead, learning nothing, failing completely to alter our course.

In summary, the purpose of this chapter was to demonstrate that money is a function of agricultural surplus. It truly *originates* from agricultural surplus, in the sense that matters most at this point in history. Agricultural surplus is what "generates" money; not tourism, not even ecotourism and certainly not the bank. Therefore, money supplies indicate the amount of agricultural surplus, and in turn the ecological footprint. Lots of agricultural surplus generates lots of money. No agricultural surplus generates no money. Limits to agricultural surplus means limits to money.

Real money, that is.

CHAPTER 8

Technological Progress and Less-Brown Growth

> *Nobody really knows why the US economy could generate 3 percent annual productivity growth before 1973 and only 1 percent afterward.*
>
> PAUL KRUGMAN

WITH THE TROPHIC THEORY of money, we have an ecologically valid perspective on where money originates. Just as milk doesn't come from the grocery store except in the shallowest of terms, neither does money come from the bank. Milk really comes from the cow, and real money comes from the agricultural surplus that frees the hands for the division of labor. If there were no agricultural surplus and you really wanted milk, your hands would be on an udder and not on your wallet.

The trophic theory of money also provides us with an ecologically valid perspective on spending, a crucial perspective with economic growth at the crossroads. It helps here to recall the fundamental identity of national income accounting: production = income = *expenditure*. With the size of the economy becoming evermore problematic it should be clear that ever-increasing expenditures in the aggregate are not the solution. Perhaps nothing stands in the way of sustainability as much as the notion that we can spend our way out of unsustainability. Examples of this notion are all too familiar. As the politically correct approach to sustainability goes, we simply need to invest in more solar panels, windmills and cotton tote bags. Then we'll have "green growth."

"Green growth" is one of the slipperiest shibboleths in recent memory. It's an oxymoron to rival "jumbo shrimp" and "old news." It's rife with corollaries, too: green jobs, green technology, green sectors. Johnny Cash singing "Forty Shades of Green" comes to mind. The mesmerizing melody and dreamy lyrics are uncanny in the current context. We live in a political economy drunk on green beer, and we need the sobriety of skepticism.

In fact, it's time to employ another portion of the color spectrum in reference to economic growth. Green sends the wrong message; "brown" is the better word. Brown more readily invokes scraped earth, hazy air, sludgy water, stained snow and a general lack of green space. Instead of green growth, we have brown bloating.

Some consumable goods are less brown than others—think Honda vs. Hummer—but even producing a unicycle requires natural resources and entails pollution. It just doesn't square to call an expanding unicycle sector a "green" phenomenon. Even compared to Hummers, unicycles are just less brown, not green. Yes, it takes a lot of iron to manufacture Hummers. Mining so much iron, and then driving Hummers, turns a lot of the Earth brown. But mining some iron for unicycles removes some green, too. The mining and then the riding of unicycles turns the Earth a shade more brown. Yes, the growth of the unicycle sector is less brown than the growth of the Hummer sector, but neither sector's growth is "green."

The service sectors, too, have their role in the browning process of economic growth. From driving trucks (quite a brown service) to answering phones (less brown, on the surface), material inputs and pollution is part of the deal. We also have to remind our green-beer-drinking friends that much of the phone answering is in service to the trucking sector. In more general terms, in the information economy growing quantities of information feed the already-brown sectors. If we don't remember this, the Green Sheen Machine will continue to get away with talk of "de-materializing" the economy, lulling citizens and policy makers into leaving environmental concerns for tomorrow, while we experiment with "greening" our growth today.

We shouldn't be surprised if they start talking about "green population growth" for green jobs and green consumerism. After all, cheaper labor and more consumers is what corporations want. So we also have to remind our green-beer guzzlers that Hummer drivers and unicycle riders alike—indeed any producer or consumer of any good or service—must be fed, clothed and sheltered. Population growth, long encouraged for the sake of economic growth and now even encouraged for the sake of increasing growth per capita (Chapter 5), entails the production and consumption of more food, clothing, shelter and the wide range of other goods and services entailed by a human life today. It's not always and everywhere bad, but it's never, nowhere green.

All the fuzzy talk about green growth stems from a simplistic view of technological progress. The fact that there is a basic conflict between economic growth and environmental protection is generally understood. By "basic conflict," however, I mean more specifically the conflict between economic growth and environmental protection in the absence of technological progress. This conflict has been described by professional, scientific societies such as the American Society of Mammalogists, the US Society for Ecological Economics and The Wildlife Society. These and several other scientific societies have adopted positions that describe a trade-off between economic growth and environmental protection. They have adopted such positions in order to raise public awareness of the conflict and to remind politicians that economic policies have major environmental consequences. These scientifically rigorous positions stand in the starkest contrast to the notions of neoclassical economists pertaining to growth and the environment.

For their part, politicians have at times acknowledged this conflict as well. For example, in the first sentence of the Endangered Species Act of 1973, the 93rd Congress of the United States stated that "various species of fish, wildlife, and plants in the United States have been rendered extinct as a consequence of economic growth and development untempered by adequate concern and conservation."[1] However, the phrase "untempered by adequate concern and conservation" left a theoretical door open for reconciling the basic conflict between economic growth and species conservation. The 93rd Congress was placing its hope in technological progress. Presumably with "adequate concern," technological progress could be steered in such a manner that economic growth and environmental protection, including species conservation, could be reconciled.

What was overlooked by Congress, and continues to be overlooked by most in the green genre, is the tight linkage between technological progress and economic growth stemming from *preexisting, clearly-brown* levels of technology. This linkage is also overlooked by most in the "neogreen" genre, that fuzzy mix of neoclassical economists, naive environmentalists and green marketeers who think we can reconcile economic growth with environmental protection. In fact, the nature of this linkage is quite possibly the single most important and widespread technical oversight in discussions of economic growth and environmental protection.[2] It's time to explore this linkage in detail, to expose it to widespread public scrutiny and to consider the implications for economic policy and consumer behavior. It will help to start with a short review of the basic conflict between economic growth and species conservation, then to focus on technological progress, finding ultimately that there is not only a basic conflict, but a *fundamental* conflict between economic growth and environmental protection.

With the basics of thermodynamics (Chapter 6) and trophic theory (Chapter 7), we were able to examine the economic production process from an ecological perspective. We saw that the foundation of the human economy is agricultural and extractive activity that directly impacts fish, wildlife, plants, and all other non-human species. In the United States, for example, agriculture, mining, logging and domestic livestock production are all prominent causes of species endangerment.[3] The production of crops, ores, logs and livestock requires the conversion of natural resources into human goods. As we saw in Chapter 6, the natural resources may be considered natural capital. Stocks of natural capital are drawn out of the economy of nature so that goods and services may flow into the human economy. In the absence of *Homo sapiens*, natural capital is allocated entirely to fish and wildlife. It's like money in the bank for biodiversity. Conversely, when humans are prominent, considerable reallocation of natural capital occurs.

If stocks of natural capital are drawn upon faster than they can be replenished, the drawdown enters the phase of liquidation. This occurs, for example, when a forest or a fishery is harvested at a rate exceeding its sustainable yield. There are numerous examples of species extinction and endangerment resulting from the liquidation of natural capital. For example, liquidation of old-growth forests in the Pacific Northwest has endangered the spotted owl[4] and, in an even more straightforward manner, liquidation of Atlantic cod stocks has endangered the cod as a species.[5]

The manufacturing sectors also entail a drawdown of natural capital because the elements of manufactured consumer goods and manufactured capital are procured or derived from nature. In 2012, for the first year in history, 60 million automobiles were set to roll off the world's assembly lines.[6] Over 600 million passenger cars alone are on the roads today.[7] Imagine the mountains of iron ore, aluminum and other raw materials mined for the chassis, drive trains and sundry parts of these cars. That's a drawdown of natural capital for only one manufacturing sector. Machinery, cement, textiles, apparel, wood products, paper, chemicals, furniture, plastics, rubber...lots of drawing down of all kinds of natural capital. Much of the natural capital being drawn from is not renewable. For nonrenewable resources such as minerals, drawdown is essentially synonymous with liquidation (although some proportion of minerals may be recycled).

Service sectors, on the other hand, are often portrayed as less dependent on natural capital.[8] We have already explored the self-sufficient services fallacy, but here we should also note that some service sectors directly and continuously require copious amounts of natural capital, especially energy feedstocks. For example, the transportation sector requires a continuous drawdown of petroleum stocks. Other service sectors such as banking, insurance and computational services appear less directly involved in natural capital drawdown, especially after the required infrastructure, buildings and equipment are in place. When such services become prominent, the economy is christened an "information economy" as described in Chapter 7.

We are almost to the point of addressing the key, overlooked aspect of technological progress, yet one more observation is important to this prelude. The role of technological progress in environmental affairs is far more complex than most topics that enter into public dialog. It will be one of the most challenging topics for the polity's intellect in the 21st century. The complexity of this issue helps to explain the popularity of a simplistic notion with a fancy title—the environmental Kuznets curve.

The environmental Kuznets curve represents the argument that there *is* a basic conflict between economic growth and environmental protection, but that the basic conflict is resolved when *enough* growth occurs (Figure 8.1). This might remind us of the game we played in Chapter 6, where more smoking led to "increas-

FIGURE 8.1. Environmental Kuznets curve: a grain of truth embedded in a fallacy.

ing supplies" of clean air, but the logic is that, when enough financial wealth accumulates, especially in per capita terms, society turns part of its focus to solving environmental problems. Why? Because environmental problems have proliferated as a result of economic growth, thereby increasing demand for environmental problem-solving. Furthermore, as a result of economic growth, society now has the money to spend on environmental problem-solving!

There is clearly a grain of truth to the environmental Kuznets curve, because the inhabitants of impoverished nations are, by definition, struggling to subsist. Little attention and less fiscal resources are available for environmental protection agencies and programs. Also, there is empirical evidence for Kuznets curves in microeconomic scenarios. For example, sulfur dioxide emissions have been reduced in nations that accumulated enough fiscal resources to ascertain the problem, develop technological alternatives and replace the problematic infrastructure.[9]

However, here we are concerned with economic growth, the macroeconomic process of increasing production and consumption of goods and services *in the aggregate*. There is no evidence for a macroeconomic environmental Kuznets curve. Many sector- or industry-specific environmental problems have been created and exacerbated as a function of economic growth; few have exhibited a Kuznets curve.[10]

The environmental Kuznets curve is especially irrelevant to biodiversity. (Conversely we might say the fallaciousness of the environmental Kuznets curve is especially evident with biodiversity.) That's because biodiversity loss is an environmental problem with a particularly macroeconomic aspect. The complete collection of species in a nation or on the planet is the "macroeconomy of nature." We might spend some of the money we generated from liquidating natural capital on saving an endangered species (a microecological accomplishment), but in the process of liquidating natural capital we endangered many others (a macroecological outcome). In more technical terms, "economic growth proceeds at the competitive exclusion of nonhuman species in the aggregate."[11]

I witnessed this principle first hand a long time ago as the Recreation and Wildlife Director for the San Carlos Apache Tribe. The biggest elk antlers in the world come from the San Carlos Apache Reservation, situated in east-central Arizona. Most of the elk hunting on the reservation is reserved for the Apaches, but some hunting permits are sold to non-Apaches for tribal revenue. During my last year with the tribe in 1993, we sold three special elk-hunting permits for $43,000 apiece, and the revenue was earmarked for elk habitat improvement. An economist would say that the elk hunting "generated" revenue for the tribe. What caught my attention was that two of these permits—$86,000 worth—were purchased by one Aaron Jones, who owned the largest old-growth sawmill in the Pacific Northwest. It was the liquidation of old-growth Douglas fir and western red cedar that really generated the money. Some of the money was then spent on elk permits in Arizona, and some elk habitat was improved. Meanwhile, however, the spotted owl had been listed as an endangered species due to the liquidation of

FIGURE 8.2. Clearcut near Corvallis, Oregon, generating money for "green" expenditures elsewhere. Credit: Alexey Voinov

ancient forests in the Northwest. Furthermore, the spotted owl was widely recognized as an "indicator species" because its plight indicated the demise of whole ecosystems and the many species therein. In other words, the net environmental effect of the elk-hunting transaction was negative because it entailed the liquidation of old-growth forests to generate the money to buy a couple of permits. That's the trophic theory of money. It trumps the supposed environmental Kuznets curve because it is ecologically and macroeconomically sound.

Almost as an aside, you might ask how we actually "improved" elk habitat at San Carlos with the revenue generated from old-growth logging. After all, most of the existing elk habitat was pretty good already, or hunters wouldn't have been flying in on Lear jets to hunt them. What we did was buy out some grazing leases from one of the tribal cattle associations, dedicating approximately 6,000 acres to elk rather than cattle production. That of course lowered the revenue from cattle, further nullifying the Kuznets curve.

Then there is the issue that, regardless of potential test-tube efforts to clone animals from preserved specimens, an extinct species cannot be resurrected to function with any semblance of ecological integrity, regardless of how much money is spent trying. Cloning makes for good movies, like *Jurassic Park*, and maybe for barnyard freak shows, but not for real ecosystems.

Similarly, it is exceedingly difficult to restore habitats that have been wholly transformed. For example, a metropolis replete with economic activity cannot be returned to a state of ecological integrity with a full complement of its original species, even if the metropolis is abandoned and re-occupied by non-human species. Furthermore, abandonment of a metropolis would contribute not to economic growth, but to economic recession, thereby nullifying the environmental Kuznets curve, which applies to the condition of economic growth. Unsurprisingly, the few studies designed to detect a biodiversity Kuznets curve have not done so.[12]

Using advanced statistical methods to investigate the relationships among population, affluence and environmental impact, scientists at Michigan State University concluded: "Contrary to the expectations of the EKC [environmental Kuznets curve], increased affluence apparently exacerbates rather than ameliorates impacts."[13] They also found that "the proportion of GDP in the service sector, the proportion of the population that is urban, and the proportion of the population in the high consumption and production age groups have no net effect on environmental impact."[14] Their findings corroborate the trophic theory of money. The real economy grows as an integrated whole with a trophic structure, even though GDP proportions mislead economists into thinking that somehow the service sectors have become far more prominent relative to agricultural, extractive and manufacturing sectors.

We can illustrate this principle with the metaphor of the "800-pound gorilla." Like the "elephant in the room," the 800-pound gorilla is often used to describe a problem that people don't want to acknowledge, even though the problem gets harder to ignore. Here the problem is the economy, in particular a growing economy. It's already an 800-pound gorilla, and it's growing. As the gorilla grows, it tends to grow as an integrated whole. We don't have a head growing while the feet shrink. We don't have service sectors proliferating while agriculture declines.

The demand for the gorilla's head (where the information resides) might increase faster than the demand for the gorilla's feet, which are dirty on the ground (like agriculture). So the proportion of total expenditure on the head may increase compared to the proportion of total expenditure on the feet, but the feet and the rest of the body have to grow to support the growth of the head. If we only look at expenditures (GDP), we might be fooled into thinking the head is growing proportionately larger than the feet, but in reality the gorilla is growing as an integrated, proportional whole.

Input-output analysis may be used to get a more technical view of the gorilla. Francois Quesnay's *Tableau Economique* was a prototypical input-output model, but Wassily Leontief (1905–1999) is credited with developing modern input-output analysis, and he won a Nobel Prize for it. Leontief used a matrix in which inputs were indicated in the columns and outputs were indicated in the rows. Such a matrix helps demonstrate how dependent each sector or industry is on the others. After all, each sector is a customer of other sectoral outputs as well as a supplier of inputs for other sectors. A column of an input-output matrix shows the value of a sector's inputs; a row shows the value of a sector's outputs. If you change a value in any cell, many or all values in the matrix will change as well, and so will the sum total of output. National income accountants use input-output analysis to "square up" the production and consumption of an economy's sectors, and in the process are able to ensure a more accurate estimate of GDP (a measure of the full gorilla).

So we see that the grain of truth in the Kuznets curve finds no fertile soil in the environment. We might save a species here or solve an environmental problem there by spending enough money, but *generating* the money endangers other species and causes other environmental problems. The feet of the 800-pound gorilla are growing, and so is the ecological footprint. The only hope for reconciling economic growth with species conservation and environmental protection lies squarely with technological progress, not some fallacious environmental Kuznets curve that was probably "generated" (and certainly circulated) with corporate funding.

So finally we come to the intellectually tougher issue of technological progress. The phrase "technological progress" connotes invention and innovation, or new technology and technological regimes. In economic terms, technological progress increases productive efficiency, or productivity; that is, a greater production of output per unit input.[15] Engineers may view such an increase in physical terms, for example, an increase in auto chassis production from the same amount of iron. Energy use may or may not be factored in, depending on the context. But economists try to level the field in assessing productive efficiency by using monetary units, such as dollars, in measuring inputs and outputs.[16] In other words, technological progress — more value produced with a given investment of dollars — is a means of increasing profits. This is a crucial point as we consider the relationship of technological progress to economic growth and thence to environmental protection.

First, though, it is necessary to distinguish among types of innovation. Conventional economists typically distinguish between product innovation and process innovation.[17] Product innovation is synonymous with invention. Coming up with the proverbial better mousetrap is an example of product innovation. Process innovation pertains to re-configuring the production process, often by using new inventions. However, in paying particular attention to the prospects for alleviating environmental impact, we may identify three categories of innovation that are more relevant to the relationship between economic growth and environmental protection: explorative, extractive and end-use.[18] Each of these categories may entail product or process innovation.

Explorative innovation allows the user to locate stocks of natural capital that were not previously detectable. For example, the magnetometer is a device used to measure the strength and direction of magnetic fields, which is useful for locating minerals, most notably iron. An early version was invented by C. F. Gauss in 1832, and reiterative innovation has led to the discovery of additional iron ore and subsequent extraction. Explorative innovation increases the production and consumption of goods and services by increasing the amount of natural capital reallocated from the economy of nature to the human economy, not by increasing the efficiency of production from known reserves.

Extractive innovation allows the user to extract known resources that were previously inaccessible. For example, helicopter logging has developed as a means to extract logs from forests that are physically or legally off limits to more conventional logging methods. Ongoing innovations in helicopter logging have led to higher rates of harvest. As with explorative innovation, extractive innovation generally contributes to economic growth by increasing the amount of natural capital reallocated from the economy of nature to the human economy. That would hardly qualify as environmental protection.

This leaves end-use innovation as the lone source of technological progress that could conceivably reconcile economic growth with environmental protection. End-use innovation is essentially synonymous with increasing productive efficiency. A good example for our purposes is the fuel efficiency of fishing vessels. Product innovation such as engine or vessel design, or process innovation such as optimizing fishing time to account for daily weather patterns, may increase the amount of fish that may be caught per unit of fuel consumed. Two basic scenarios may follow. In the first scenario, the same amount of fish are caught and sold, but less fuel is purchased. All else equal, economic growth does not result. In the second scenario, the same amount of fuel is purchased, more fish are caught and sold and all else equal, economic growth does result.

In the first scenario, economic growth is not reconciled with environmental protection because economic growth does not occur. In the second scenario, economic growth is not reconciled with environmental protection because, although economic growth occurs, more fish are reallocated from the economy of nature to the human economy.

Yet we cannot quite conclude that end-use innovation cannot *possibly* reconcile the conflict between economic growth and environmental protection, because there is a great deal of nuance in the market that negates the phrase "all else equal." (In economics jargon, *ceteris paribus* means all else equal.)

For example, lower demand reduces the cost of fuel, which may then be used more liberally in other economic sectors, impacting the environment in other ways. In yet another scenario, somewhat less fuel is purchased by the fishing fleet, and somewhat more fish are caught and sold. This is an intermediate scenario compared to the two basic scenarios described above. Theoretically, the economy may grow somewhat, with somewhat less natural capital reallocated from the economy of nature to the human economy.

These theoretical considerations do nothing to quantify the productive efficiency gains required to reconcile the conflict between economic growth and environmental protection. However, the

Michigan State researchers mentioned above estimated "an annual growth rate in the global footprint of 2.12% per year. If, as frequently suggested, technological progress can redress environmental problems (Ausubel 1996), the requisite technological improvement needs to exceed 2% per year."[19] Productivity gains exceeding two percent per year typified the "advanced capitalist economies" during the third quarter of the 20th century,[20] but gains falling well below two percent per year have befuddled growth theorists, national income accountants, and Paul Krugman ever since.[21] Unfortunately, macroeconomic efficiency data are largely unavailable for many nations and notoriously difficult to compare, but a number of studies suggest that productivity gains have been and continue to be lower in the vast majority of countries not categorized as advanced capitalist economies, even among the "Asian Tigers."[22] For example, Malaysia's GDP increased at a rate of 6.48 percent per year from 1980–2001, while Malaysian total factor productivity (a standard measure of productive efficiency) increased by only 1.29 percent per year. In other words, much of Malaysia's economic growth resulted from an increase in factor inputs (land, labor and capital), and not from the efficiency with which those factors were used. Similar scenarios throughout Southeast Asia, where deforestation is occurring more rapidly than in any other major tropical region, have led to an "impending disaster" of biodiversity loss.[23]

The only solid conclusion to be drawn thus far is that, in theory, the *possibility* of reconciling economic growth with environmental protection via technological progress, at least temporarily, cannot be denied. However, it is not occurring and does not appear likely, partly because much innovation is not end-use but rather explorative and extractive, both of which increase the reallocation of natural capital to the human economy. Even end-use innovation does not appear to offer a sure prospect for reconciling substantial rates of economic growth with environmental protection, although it does appear to offer a clear prospect of lessening the impact of economic growth (at any rate), and perhaps a lesser prospect of reconciling modest rates of economic growth with some aspects of environmental protection. Any of these prospects are eventually limited by the entropy law, which establishes limits to productive efficiency (Chapter 6). In other words, any prospective reconciliation of economic growth with environmental protection via end-use innovation would have to be viewed as temporary in nature.

So far in this chapter, technological progress has been described in a simplistic context reminiscent of older models of economic growth in which technology was characterized as "manna from heaven" (Chapter 5). The Romer model helped to illuminate the more complex (as opposed to miraculous) relationship between economic growth and technological progress. However, in the hands of Romer and his neoclassical colleagues, even this "endogenous growth theory" didn't shed the necessary light on the prospects for reconciling economic growth with environmental protection. Recall that the quintessential neoclassical conclusion is that increasing human population is required for increasing per capita consumption. So let's back up a bit and reconsider the nature of technological progress as it relates to economic growth *and* to environmental protection.

As Romer pointed out, technological progress stems from research and development, or R&D.[24] R&D is conducted by industry, governments, colleges and universities and non-profit organizations. The US (including government and non-governmental sources) invests approximately $300 billion per year in R&D, or approximately 30 percent of global R&D[25] and 41 percent of the R&D conducted by nations comprising the Organization for Economic Cooperation and Development.[26] Only Israel, Sweden, Finland, Japan and Iceland, in that order, invest higher percentages of their GDPs than the US. Given the leading role played by the American economy and technology, let us scrutinize the American data to gain insights about the nature of R&D with regard to productive efficiency and therefore the prospects for environmental protection.

In the US, about 71 percent of R&D is conducted by industry, 17 percent by colleges and universities, 7 percent by the federal government, and 5 percent by non-profits.[27] Similar proportions apply in other nations with substantial R&D expenditures. These figures refer to who conducts the R&D, not the funding source. For example, the US government provides funding for R&D conducted by industry, colleges and universities and non-profits, both directly (e.g., through grants and contracts) and through the administration of research and development centers. Conversely, colleges and universities receive most of their funding for R&D from federal and state governments, industry and non-profits.

Whether viewed in terms of who is conducting the R&D or who is providing the funds, the R&D landscape is dominated not only by the US but by corporations, both within and outside the US. Of the world's 100 largest economic units in 2000, 51 were corporations and 49 were countries.[28] Corporations are chartered for the primary purpose of generating profits and seldom deviate from that purpose in their investment and management decisions. Consequently, the vast share of global R&D is conducted for the purpose of generating profits, not for the purpose of environmental protection. This observation is not intended to reflect the level of environmental concern among corporate executives, personnel and shareholders but rather the financial exigencies of corporate survival. The onus is less on corporations to invest in technologies conducive to environmental protection than on citizens and policy makers to insist upon environmental protection, via public policy when necessary.

One would be tempted to assume that any R&D leading to greater productive efficiency has the potential to reconcile the basic conflict between economic growth and environmental protection, as long as enough such R&D is conducted. However, as described above, much of technological innovation is explorative and extractive and tends to increase the drawdown of natural capital, whether or not productive or financial efficiency increases. The National Science Foundation does not categorize R&D as explorative, extractive and end-use innovation, but the categories it does use provide clues about the nature of R&D expenditures. "Basic research"

Technological Progress and Less-Brown Growth | 211

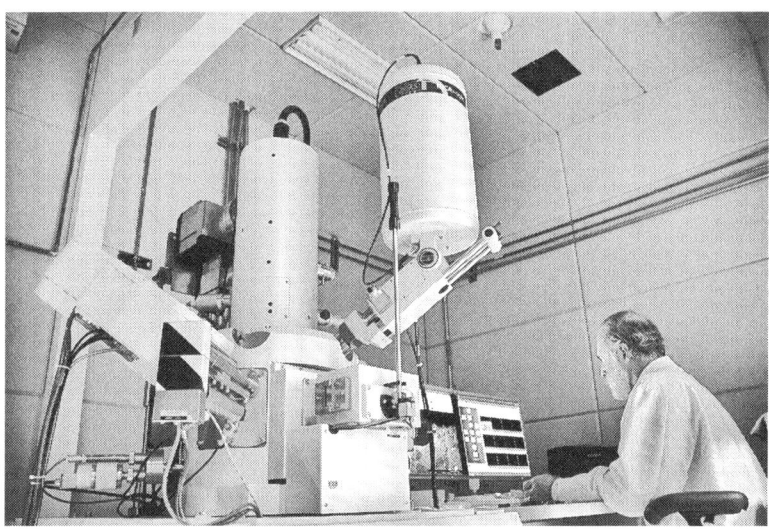

FIGURE 8.3. Research and development: the key to technological progress and economic growth. A scientist conducting basic research at Idaho National Laboratory. Credit: Idaho National Laboratory

FIGURE 8.4. R&D has a direct and obvious environmental impact, as with the Ames Research Center at Moffett Field, California. Less obvious is that R&D requires a far bigger ecological footprint, dispersed throughout the planet, to generate the money to fund expensive R&D facilities and programs. Credit: NASA

connotes scientific investigation for the general benefit of society. "Applied research" is designed to answer specific questions for particular users. "Development" refers to the innovations required to bring the answers into practice. US R&D consists of approximately 19 percent basic research, 23 percent applied research, and 58 percent development. Given the corporate focus on profits, it is not surprising that the lowest proportion of basic research (4 percent) and the highest proportion of development (74 percent) is conducted by corporations.[29] "Development" in this context means bringing products or processes online that will increase profits, after the products or processes are identified with applied research. Conversely, the highest proportion of basic research (70 percent) and the lowest proportion of development (6 percent) is conducted by colleges and universities.

Trends in recent decades have been away from basic research conducted by universities and governments and toward development conducted by industry. For example, regarding the "very substantial investments in agricultural research," Vernon Ruttan, a prominent scholar of science and technology, noted: "Initially this research was conducted primarily in public sector institutions—experiment stations and laboratories operated by ministries of agriculture or universities. Since the 1970s private sector research organizations operated by seed companies, animal breeders, and chemical companies have come to account for a larger share of agricultural research directed specifically to development of technology."[30]

Furthermore, the profit motive (in the corporate sphere) and the macroeconomic goal of growth (in the government sphere) tends to redirect savings from any end-use innovation toward other activities that increase production and consumption in the aggregate. Such activities include capital investments to increase production, marketing to increase consumption and further investments in R&D, including explorative and extractive R&D and especially in the development phase. This leads to a closer consideration of the linkage between economic growth *at current levels of technology* and the technological progress that raises the ceiling for further economic growth. We are getting, in other words, to the crux of the matter.

First, however, we need to get an idea of the scale of R&D that may be motivated to some degree out of concern for environmental protection. Approximately $14 billion of R&D was expended on earth and biological sciences at US colleges and universities in 2006.[31] Some of this was probably designed to also increase production, as with much of the nearly $3 billion of agricultural R&D. Presumably some of the $9 billion of R&D classified by the National Science Foundation as "biological" was oriented toward environmental protection. Even if a third of that amount—an extremely generous estimate—was oriented toward environmental protection, it would comprise only one percent of US R&D.

Some government R&D may be focused on environmental protection, but whatever the amount is dwarfed by R&D devoted to defense in nations such as the US, China and the Russian Federation.[32] In Japan, where military objectives are limited, approximately 35 percent of R&D is devoted to "economic objectives"[33] (a somewhat euphemistic phrase referring in essence to economic growth). In China, the suddenly developed superpower with the second-highest R&D expenditures, most R&D is devoted to defense and economic growth.[34]

It is difficult to obtain data on the R&D expenditures of non-profit organizations, but presumably non-profit R&D budgets reflect broader societal trends and concerns, as well as funding sources. Nearly half of the funding for US non-profit R&D is provided by the federal government (39 percent) and industry (9 percent) while, somewhat paradoxically, the rest is provided by other non-profits.[35] Of the 100 largest non-profits in the US, only five could be described as devoted to environmental protection: The Nature Conservancy (20th largest), Wildlife Conservation Society (52), Ducks Unlimited (55), American Museum of Natural History

(76) and the Trust for Public Land (95).³⁶ These organizations, like most environmental non-profits, are focused on applied environmental conservation mixed with small amounts of research.

This paucity of environmental conservation R&D relative to military or growth-oriented R&D is but one aspect of R&D that lessens the prospect of reconciling economic growth with environmental protection. Another aspect is that most environmental R&D produces policy implications conducive to tempering economic projects or activities rather than conducing economic growth, so that even environmental R&D in the aggregate cannot be assumed to contribute to reconciling (whether or not that is possible) the basic conflict between economic growth and environmental protection. R&D designed to increase productive efficiency in various economic sectors appears to have more potential to reconcile the basic conflict, while R&D designed to ascertain the economic causes of environmental degradation has more potential to raise awareness of the conflict.

By definition the 65 percent of R&D coming from corporations must be a direct function of corporate profit. It is only after the factors of production (land, labor and capital) are paid for and shareholder dividends allocated that any remaining corporate revenue may be allocated to R&D. A similar requirement applies to the next-largest funder, the federal government, in the sense that its revenue comes almost entirely from income taxes (individual and corporate) and social security payments. Profits and individual incomes are tightly linked because most individuals draw their wages from profit-making firms.³⁷

Taxes and social security payments are forthcoming only from solvent firms and individuals; that is, those that have paid off maintenance and subsistence bills, respectively. In other words, federally funded R&D is a function of profits. Colleges and universities, non-profits and state and tribal governments are also dependent upon profits or, in more general terms, income above maintenance and subsistence costs. "Ceteris paribus," increasing R&D requires increasing profits at the corporate level and increasing income at the national level—that is, economic growth.

The tight linkage of R&D to economic growth, and vice versa, is indicated by the fact that R&D itself comprises a distinct category of expenditure in the calculation of GDP. This reciprocal linkage was illuminated by the Panel on Research and Development Statistics at the National Science Foundation: "The [R&D expenditure] data are sometimes used to measure the output of R&D, when, in reality, in measuring expenditures, they reflect only one of the inputs to innovation and economic growth."[38] Likewise, in neoclassical growth theory, R&D and economic growth are modeled as mutually reinforcing processes. In fact, neoclassical growth theory is often applied to promoting specific R&D programs on the grounds that R&D will contribute to the goal of GDP growth, such as in the aerospace industry.[39]

Vernon Ruttan captured the reciprocal relationship between economic growth and R&D while providing insights about the material nature of an information economy:

> The most important and visible output of a research laboratory is the *information*, in the form of new knowledge or new technology. At the more fundamental or basic end of the research spectrum, the new knowledge may be embodied in published research papers. At the technology development end of the spectrum the research may result in patent applications. The ultimate test, however, is whether the new knowledge is embodied in a new product or a new practice.... If a research institution or system is to achieve economic viability, the flow of new knowledge and technology that it generates must in turn generate new income streams. These new income streams may accrue largely to the sponsoring organization in the case of privately funded research, or to society more broadly in the case of publicly funded research.

If a research system is to remain a valuable private or social asset, it must also devote resources to reinvestment in institutional capacity—to the enlargement of its own physical and intellectual capital. This means diverting some resources to the production of information that does not have immediate application. This also means expanding the capacity of its scientific staff through time devoted to graduate education, study leaves, and supporting of basic research. The facilities, administrative structure, and ideology that serve as a rationale for the research program must also be continuously updated in response to new scientific and technical opportunities, changes in the market environment, or changes in social priorities.[40]

These basic relationships among profits, R&D, technological progress and economic growth must be grasped by scientists, citizens and policy makers who wish to add value to scholarly or public dialog about the prospects for reconciling economic growth with environmental protection. Most of the value added by scientists, however, will come from their knowledge of the natural sciences, because economists who have dominated discussions about economic growth already grasp the significance of the economic factors. Economists of all ilks, from the classical economists and Marxists of the 19th century to the neoclassical economists of today, have long agreed that profits dry up for the firm that fails to attain a competitive advantage. Similarly in macroeconomics, the tendency for national income to stagnate has played a prominent role in economic growth theory since Keynes wrote *The General Theory* (Chapter 4). The conventional solution offered in both cases (firm and nation; micro- and macroeconomics) is technological progress, which allows the firm to gain an advantage over its competitors and the economy to grow continually. In other words, R&D is required to maintain profits—and economic growth—while profits are required to maintain R&D.

This reciprocal requirement may be viewed alternatively as a virtuous spiral or an impossible Catch-22. With a few notable exceptions such as E. F. Schumacher and Herman Daly, economists have viewed it as a virtuous spiral. Recalling the background in trophic levels and thermodynamics provided in previous chapters, a reasonable view for the ecologically informed would be of a largely beneficial spiral until limits to growth were approached. The ecologist might hasten to add that biodiversity was continuously lost along the way, and the ecological economist would add that, long before limits to growth were breached, the growth of the economy had actually become uneconomic, causing more problems than it solved. It was "good growing gone bad," as explored in Chapter 2.

It is important to fully grasp one other economic phenomenon to understand what has happened with economic growth and technological progress vis-à-vis environmental protection. Much of the ecological literature, most notably that on petroleum supplies and the ecological footprint, indicates that the global economy is already beyond its long-run limits to growth. An economy this large is living on time borrowed from the liquidation of various natural capital stocks and funds. However, national and global economies are still growing in the short term, and not entirely as a function of technological progress. There has been one other primary source of increased productive efficiency and attendant profits. That overlooked source is called "economies of scale."

Economies of scale are "reductions in the average cost of a product in the long run, resulting from an expanded level of output."[41] They are classified as internal or external. Internal economies of scale operate within the firm, such as when the efficiency of a sawmill increases as a result of a higher rate of timber moving through the mill. External economies of scale operate at broader scales, such as when the timber industry grows large enough to hire a public relations firm that provides advertising services at a cheaper rate than would be paid by individual timber companies.

Economies of scale also operate macroeconomically. The prolific national income accountant, Edward F. Denison, attributed the increased productivity (indicated by increased income per capita) of Western European nations from the period 1950–1962 largely to economies of scale.[42] Walt Whitman Rostow helped interpret Denisons' findings thusly: "In the vocabulary of the stages of economic growth, Western Europe came—belatedly but fully—into the stage of high mass-consumption."[43]

Economies of scale arise independently of new technology. For example, increasing demand as a function of population growth induces economies of scale. In a sense, economies of scale comprise a blunt form of process innovation and, as with technological progress resulting from R&D, economies of scale do result in increasing productive efficiency. However, unlike the finer aspects of technological progress such as invention, by definition economies of scale require a concomitant increase in aggregate production. Furthermore, the financial aspect of efficiency is emphasized with economies of scale, whereas physical efficiency is more likely to be emphasized with technological progress resulting from R&D.

Combining economies of scale with R&D has clearly been the modus operandi of the big corporations because of the dual benefits of increased efficiency and market share.[44] However, this combination forms another spiral that, for purposes of almost any environmental problem, has not been at all virtuous. Economies of scale have contributed substantially to corporate profits, but they have done so by increasing the reallocation of natural capital at current levels of technology and, in the process, polluting ecosystems at increased rates. Increasing levels of production using current levels of technology cannot reconcile the basic, already existing conflict between economic growth and environmental protection. The best that can be hoped for is that the additional production attributable to new technology stemming from R&D has a lesser impact *per unit* than the impact of production based upon old technology. Less-brown growth, in other words. To the extent that R&D is financed from profits generated via economies of scale (as opposed to new technology), even that prospect is diminished.

If you've made it this far, you're beyond the most technically challenging aspects of the book. At this point it should prove helpful to clarify some terms. For purposes of informing public policy debates with regard to the relationships among economic growth, technological progress and environmental protection, we should exercise caution to avoid misleading others and to avoid being quoted out of context by individuals or organizations that promote economic growth. The common claim that economic growth can be reconciled with environmental protection given enough technological progress is scientifically sound only if "reconciled" means that the *rate* of environmental deterioration decreases as the economy grows via technological progress, not that the deterioration ceases or that the environment is somehow improved. However, "reconcile" tends to connote a thorough resolution to a problem, so "lessen" would be more apt. The argument would then take the form, "The basic conflict between economic growth and environmental protection may be lessened with technological progress."

Also, the phrase "*may* be lessened" is important, as opposed to "*is* lessened," because of the preponderance of R&D historically and currently devoted not to conservation purposes but rather to increasing profits (at the corporate level) and economic growth (at the government level). Theoretically, though, the conflict may be lessened given a particularly focused program of R&D in which research for solving environmental problems is prioritized over the development of products for profit. Among other things, this entails prioritizing end-use innovation over explorative and extractive innovation.

Also, because "basic" was used to describe the conflict in the absence of technological progress, and given that technological progress does not reconcile but at best may only lessen the conflict, "fundamental" more accurately describes the conflict. "Basic" tends to connote simple or even simplistic, whereas "fundamental" indicates that the conflict is founded in first principles and its nature is congruent with a rigorous analysis of the evidence. Furthermore, because the nature of R&D tends to reflect profit motives and the political economy of growth, additional information should be provided to clarify to the public and policy makers that there must be

FIGURE 8.5. Oil from the Deepwater Horizon spill approaching New Orleans, Louisiana, May 24, 2010. The oil appears light in color because it smoothes the ocean surface. The Louisiana coast is also vulnerable to sea-level rise, exacerbated by land subsidence (due primarily to natural gas, oil and water withdrawal). Credit: NASA

a systematic approach to lessening the impact of economic growth via technological progress. For example, the argument may take the form, "The fundamental conflict between economic growth and environmental protection may be lessened with technologies that increase productive efficiency, but this type of technological progress requires R&D policy goals and tools that are conducive to increasing productive efficiency rather than exploration and extraction."

Of course, the fact that technological progress may, theoretically, lessen the rate of environmental impact caused by economic growth, given sufficient R&D policy development, is a far cry from reality. BP oil spills, Keystone pipelines, Fukushima meltdowns... the reality is that GDP will get browner *by the unit* as well as in the aggregate when all the stops are pulled out for growth. We might call this the principle of increasing marginal brownness. (Synonymously, we might call it the principle of diminishing marginal greenness.) Technological progress can't turn the process of economic growth green. As they say, you can't make a silk purse from a sow's ear.

We are nearly ready to consider the political prospects and policy tools for a steady state economy in which the browning process has ceased and the remaining shades of green are retained. However, the nature of technological progress as described above does raise a tough question that might be disturbing to some: If technological progress is linked at the hip with economic growth at current levels of technology, then what happens to technology in a steady state economy?

This is a question that must be acknowledged, but only the future can provide a thorough answer. If I had to venture a guess, I'd say that technological progress will emerge in a steady state economy, but at a far lesser pace than what we've experienced since the Industrial Revolution. Presumably, the rate of technological progress will be more evolutionary than revolutionary. Economic growth is at the crossroads, and so are the vast institutions of R&D. By definition, in a steady state economy, R&D as we've known it will cease to be an ever-expanding engine of economic growth, and vice versa. However, inventions and innovation will arise organically, just as they did for the millions of years of hominid existence prior to the mass production of R&D.[45] Technologies will wax and wane as a steady state economy fluctuates within the capacity of the planet. Presumably the key will be to flush out the old technologies as quickly as better ones come online, keeping the economy stable, rather than attempting to produce as much as possible from as many technologies as we can possibly retain and market.

For those who are thrilled by the rush of R&D, perhaps it will serve as some consolation to recognize that ecological economics is itself a form of R&D, albeit with a heavy dose of social science. Ecological economics research (R) has helped us to better understand the limits to economic growth and the related conflict between economic growth and environmental protection. That's primarily a matter of the natural sciences, especially physics and ecology. Ecological economics research has also helped us develop the concept of uneconomic growth and to identify a safe and sustainable alternative. Subsequent development (D) entails the achievement of that alternative—that is, the steady state economy. Insights from political science, psychology and sociology may help in that regard. For many students and scholars, this type of R&D, which deals with living, breathing, thinking, socializing, evolving individuals and institutions is far more fascinating than materials science and engineering.

With economic growth threatening the environment, the economy, national security and international stability, the prospect of a steady state economy seems thrilling in its own right A lesser rate of old-style, materials-based R&D would be a small price to pay for the kids' and grandkids' future.

PART 4

POLITICS AND POLICY: THE HORSE BEFORE THE CART

CHAPTER 9

"What Have You Done for Growth Today?"

> *I believe we can grow the economy and improve the environment, and so does our Vice President.*
> WILLIAM JEFFERSON CLINTON

THE POLITICAL HISTORY of economic growth, dry as it may sound, is an important topic for the sustainability thinker, not to mention the responsible citizen concerned about the grandkids. Political history provides a feel for how "the system" works. It helps us understand why economic growth has become a primary, perennial and bipartisan goal of the American public and polity. It shows us what we are up against, sobering us for the task ahead. On the other hand, it should encourage us to find that economic growth has gone through episodes of serious scrutiny at the highest levels of government.

Despite the fact that economic growth is currently the priority in American domestic politics and policy, as well as in many other nations, history tells us there is nothing sacred about it. Economic growth may be called into question by national governments at any time, if only the people express their concerns.

For those of us who want to express our concerns, knowing something of the history of growth politics will help. After all, putting a halt to economic bloating requires not only personal action (such as conscientious consumption) but political action (such as meeting with politicians, writing letters to editors and speaking out in public meetings). It helps establish credibility in a public meeting to state, for example, "Economic growth wasn't always the number one goal. Republicans such as President Eisenhower and Democrats such as President Carter questioned whether growth should be a goal at all." Such an observation perks up the politician and the pollster. They correctly conclude, "This is a person who knows about economic growth, pays attention to politics and isn't afraid to report the findings in public." They will listen to the rest of your comment.

So let us consider the history of economic growth politics. Our focus will be on the United States since World War II, recognizing that most "developed world" politics ran parallel. The rest of the world — so-called "undeveloped" and "developing" countries — has been dealing with the aftermath of these politics.

The academic development of economics we explored in Part 2 was not confined to an ivory tower. We saw how, dating back to the physiocrats, economists were often involved in debates involving the politics and policies of nations. For a while, from the end of the 19th to the early 20th century, economists in the Western capitalist nations took a lower profile. This was partly a reaction to the effects of Marx and Marxism. As during the French Revolution 100 years earlier, revolutionary spirit in the hands of the masses was an extremely dangerous phenomenon. Marx himself mixed with a rough crowd, even in the context of an age in which political brutality was common. For example, when the International Workingmen's Association met, the bearish Marx met his match in the brutish Mikhail Bakunin. Marx's impeccable scholarship meant little to anarchists like Bakunin, much less to the proletariat. They were all for revolution and expropriating from the expropriators; beyond that, few were interested in matters of political economy. Yet many of the ruffians cited Marx as their intellectual leader. As for Marx, the abuse of his ideas led him to remark, late in life, "I am not a Marxist."

Marx's peers in political economy watched in horror as the Russian Revolution unfolded. They shied away from the big issues and resigned themselves to developing the techniques of marginal analysis, microeconomics and general equilibrium. This period of relative diffidence did not last long, however, especially in the United States, where economics as an academic discipline was borne of politics and policy. The American apologists for the economic elite had not only Marxism to contend with but Henry George. More importantly, however, came two world wars, a stock market crash and the Great Depression. The result was greater involvement of economists in government than ever before, owing to the Keynesian Revolution. The focus of all this new involvement was unemployment, inflation and economic growth.

Macroeconomics was a new role for government, and a new role for economists in government. Therefore, nothing was assumed except that high rates of employment, low rates of inflation and a decent standard of living throughout society were desirable. The question of how to achieve these aims was a matter of great debate, however. Economic growth was one of the primary candidates for keeping employment up, but was also feared for its inflationary effect. It was also known among deeper thinkers to challenge the social stability of communities. As Robert Collins, a prominent historian of US public policy and political economy, observes in *More*, his history of the postwar politics of economic growth: "Even in hard times attitudes were colored by both the promise of what growth would do *for* a community and the realization of what it could do *to* a community."[1]

A good example was the "Agrarians," a small band of scholars, poets and novelists from the American South. They were critical of the rapid cultural changes wrought by economic growth. They viewed the industrial economy as a "Prussianized state which is organized strictly for war and can never consent to peace."[2] The Agrarians' observation was prescient, for the German economy was finally recovering from World War I and Hitler would soon be implementing *Mein Kampf* with a war machine of unprecedented industrialization. Yet the "Prussianized state" was part of a much longer trend, reaching far back into pre-industrial times, in which economic growth and armed conflict were connected.[3]

The Agrarians also exemplified a political phenomenon that was to characterize the economic growth arena from then to the present. Pointing out the perils of economic growth, they were a voice in the wilderness. Their collective intellect was highly respected, yet they were marginalized from mainstream politics. Politicians were becoming increasingly dependent upon campaign machinery greased by corporate money. Meanwhile the masses tended to know little of dissenters such as Agrarians. In difficult times the masses were too busy making ends meet. In easier times they tended to use their wealth for the sake of pleasure, display and, for the more financially driven, investment. Investing in the stock market was all the rage in the Roaring Twenties. Very few learned of the arguments of the Agrarians and succeeding growth critics.

The Great Depression brought mixed results for the politics of economic growth. On one hand, it cast doubt on the prospects for further economic growth, as exemplified by the somber observations of President Franklin Delano Roosevelt:

> Our last frontier has long since been reached, and there is practically no more free land.... We are not able to invite the immigration from Europe to share our endless plenty. We are now providing a drab living for our own people.... Clearly, all this calls for a re-appraisal of values. A mere builder of more industrial plants, a creator of more railroad systems, an organizer of more corporations, is as likely to be a danger as a help.... Our task now is not discovery or exploitation of natural resources, or necessarily producing more goods. It is the soberer, less dramatic business of administering resources and plants already in hand.[4]

As incredible as it may seem to Americans today, there even were industries that favored a stable economy over a growing economy, at least temporarily. These included the railroad, petroleum and automobile industries. As Robert Collins described it: "Their motivation was complex, and undoubtedly colored by commercial self-interest and oligopolistic tradition, but it was also the consequence of a long and psychologically grueling depression."[5] These industries feared that attempts to stimulate the economy would simply result in bigger gluts, which were widely recognized as a primary cause of the Great Depression.

Roosevelt's Republican opponents throughout the 1930s established a rhetorical framework in which a belief in and pursuit of economic growth was "optimistic," "young" and indicative of "progress." They portrayed Roosevelt's economic philosophy as pessimistic and accused the pessimists of pushing an old, declining economy on the public. The "pessimists," on the other hand, saw themselves—and the economy—as "mature." They tried to foster an economy of balance, recovery and security.

However, the Great Depression was a desperate time calling for desperate measures and it was abundantly clear that some economic growth was needed, if only for the duration of a recovery. So while Roosevelt and many of his advisors felt the American economy was at the limits of growth, they nevertheless found it politically necessary to develop policies and programs to facilitate growth. This is the point at which Keynes made his longest stride onto the stage of history. Keynes's *General Theory* seemed to explain the Great Depression better than any other. Keynes also offered a way out, and Keynesian thought permeated governments throughout the West, especially in the United States and Great Britain.

Roosevelt's New Deal was the prototypical, quintessential application of Keynesian economics. Collins called the New Deal "state capitalism,"[6] the highlight of which was an extensive public works component. Federal agencies were created, reoriented or reorganized to employ people and facilitate economic growth. For example, the Tennessee Valley Authority was created to develop hydroelectric power, the Bureau of Reclamation was reoriented from irrigation to general infrastructure development and the

Reconstruction Finance Corporation was reorganized to finance a wide array of capital improvements.

We will never know how full or fast the recovery would have been in the absence of World War II. Nothing clears the markets and necessitates a fresh economic start like a major war, and no war was ever more major than World War II. Ironically, the war Keynes had tried to prevent with *Economic Consequences of the Peace* became the ultimate platform for Keynesian policy, because World War II engaged all the world's powerful governments in the management of their economies.

A nation united in war is like a massive public works program, employing men and women by the millions and requiring unprecedented production, consumption and capital investment. In the United States, mobilization of the economy for war was called the "Victory Program," and the United States became the "arsenal of democracy," producing, for example, nearly 100,000 tanks, over 2 million military trucks and almost 3 million machine guns.[7]

World War II had three major effects on the global politics of economic growth. First, in the victorious industrialized nations, especially the United States, there was a newly optimistic outlook on the prospects for growth. This outlook was greatly enhanced by the jubilation over winning the war. Not only were the prospects bright, but now the results of growth were viewed more positively than ever. After all, economic growth was part of the program for defeating Hitler, Mussolini and Hirohito. It is easy to see that, on the heels of the Great Depression and World War II, the pursuit of growth became much more politically powerful than ever before. But headier days were yet to come.

The second major effect of World War II was the Cold War. The Western capitalist nations were not the only victors of World War II whose economies were supercharged by the war. Russia's economy had long been riven by the likes of Mongolian invasions, aristocratic czars and eventually the Russian Revolution. The Soviet Union was formed in 1922 and struggled through several disastrous economic plans. It wasn't until World War II that the industrial revolution came fully to the Soviet Union, and the communist government proved a worthy host for industry during the war and for some time after. The Soviet Union emerged from the war as the world's second superpower, and the Cold War became a contest of military muscle building. Given World War II's lesson, however, that the power of a war machine would mirror the size of the economy that produced it, another way to keep score was with GNP. This was also an easier way, because comparing military strength is a highly subjective and somewhat clandestine endeavor, while comparing GNP and other macroeconomic parameters is relatively straightforward.

Furthermore, the Cold War wasn't *only* about military might. It was a contest of ideology, political economy and geopolitical pride. It was Adam Smith vs. Karl Marx, capitalism vs. communism, and even the West vs. the East, especially after the 1949 communist revolution in China. A wide variety of performance measures would be monitored, but most would be macroeconomic, including unemployment, inflation, wages, capital accumulation and labor productivity. The key indicator, however, was GNP. While the historian Russell Weigley called World War II a "gross national product war," the description is more fitting yet for the Cold War. In fact, the linkage between economic growth and Cold War victory was firmly and formally established in a 1950 report of the US National Security Council to President Truman.[8] This famous document, NSC-68, concluded: "In summary, we must, by means of a rapid and sustained build-up of the political, economic, and military strength of the free world, and by means of an affirmative program intended to wrest the initiative from the Soviet Union, confront it with convincing evidence of the determination and ability of the free world to frustrate the Kremlin design of a world dominated by its will."[9]

The third major effect was the establishment of international institutions that would promote economic growth, most notably the World Bank and the International Monetary Fund (IMF). They were the brainchildren of a distinguished group of economists who gathered at Bretton Woods, New Hampshire, in July 1944. Keynes was there as the major contrarian to laissez faire. Thereafter, not only would nations be heavily involved in the management of their economies, but there would be prominent, quasi-governmental agencies operating on a global scale to manipulate exchange rates, trade and credit.

The primary goals of the Bretton Woods regime were currency stabilization, reconstruction of shattered economies and the international integration of economies. The economists at Bretton Woods thought these goals would contribute to a more peaceful political economy and the Marshall Plan was drafted largely to support this agenda. However, the effects of Bretton Woods on global political economy were overshadowed for decades by the Cold War. Meanwhile the World Bank and IMF devolved into growth-at-all-costs institutions as described by John Perkins in *The Economic Hit Man*.

To summarize, the effect of World War II on economic growth, as well as its effects on the politics of economic growth and the policies supporting it, was positive, strong and reinforcing. This was one of the most important outcomes of the war, and *the* most important today, with economic growth at the crossroads. It reoriented many governments toward the goal of economic growth. This in turn funneled tremendous resources into the economics profession, resources that demanded that government and academia focus on economic growth. There was very little debate about when or even whether economic growth could become a bad thing and an inappropriate goal—even as the world was already full enough to have virtually all of its nations engaged simultaneously in war. There easily could have been such debate, with the Nazi demand for "lebensraum" ("living space" for economic growth; the rationale for invading Poland) so fresh in memory. Yet it was as if the nations, especially the superpowers, were horses in a mad race, wearing the blinders of ideology, paying no heed to the perils awaiting at the finish line. To be more specific, national *economies* were the horses, pulling the war wagons of the Cold War. Even after the

Soviet disintegration, the American warhorse would find it hard to stop running. Arab-Israeli conflicts, the Korean nuclear threat and a general "war on terror" in response to 9/11—these and a host of other full-world problems keep the old horse running to this day. The prospects for a well-deserved rest—without losing the race of national security—will be explored in Chapter 11.

For a while after World War II, economic growth came easily to most nations. For Germany and Japan, the former axis powers, the starting points were lower and the startup phases longer, burdened as they were by international obligations. They would eventually catch up with a vengeance, most famously Japan. Allied nations generally had it easier from the start, although most of the European nations were heavily burdened with repairs at the outset. The Soviet Union also had massive damage to contend with, especially to its labor force. It also took on the complicated political task of puppeteering the governments of Poland, Ukraine and other Slavic neighbors while colonizing Asian countries with unfamiliar cultures. It wouldn't be long, however, before these "republics" contributed substantially to Soviet GNP.

Among the post-war nations, the United States was the most advantaged, having sustained no physical damage within its borders (except in Pearl Harbor) while having developed tremendous capacity. Not wanting to lose this capacity, in 1946 the United States passed the Employment Act, which committed the federal government to promoting "maximum employment, production, and purchasing power."[10] The act established the Council of Economic Advisors (CEA), consisting of three members. The first CEA consisted of J.D. Clark (the son of John Bates Clark), Edwin Nourse and Leon H. Keyserling.

The CEA rapidly identified economic growth as the primary economic goal of the United States.[11] At first, it seemed to support a general redistribution of wealth, but within a few years took the position that economic growth would automatically alleviate poverty across the board. This was an early manifestation of the "trickle-down" theory that would come to characterize Republican politics four decades later, also popularly expressed as "a rising tide lifts all boats."

Keyserling called the CEA's emphasis on growth economics "the one really new thing" in economic policy.[12] By that time in American history, the Democratic Party, led by Harry Truman, had become the champion of economic growth. In a temporary reversal of roles, the post-war Republican Party was more concerned with economic stability, while Democrats had been mounting large spending programs ever since the New Deal. Big government was called for by *The General Theory*, and the Democrats subscribed. That's how they got their reputation as "deficit-spending liberals," a reputation that to some degree countered the popularity of the spending programs they fostered. But if economic growth itself were accepted as a policy goal, then Keynesian "stimulus spending" could be portrayed as contributing to this larger goal.

In 1947, the CEA began making five- to ten-year projections of GNP, and by the fall of 1949, President Truman set a precedent for using GNP as an indicator of national success. He identified a $300-billion economy as a benchmark, along with a doubling of income for a typical family.[13] This theme would be echoed in presidential politics many times, as recently as 1996.[14]

During Truman's presidency, corporations and labor unions also turned more decisively pro-growth. General Motors and the United Auto Workers signed a contract ensuring an increasing real wage. Growth expectations were running high, to put it mildly, as the "Treaty of Detroit" ensured the auto workers a 20 percent increase in their standard of living over a five-year period.

Congress, too, participated in the reorientation of American politics toward growth. The ambitious Economic Expansion Bill of 1949 stalled, but the Defense Production Act of 1950 linked national security to economic growth. The Defense Production Act was a natural outcome of NSC-68, and was further motivated by the outbreak of the Korean War, which added to the Cold War demand for military production. This demand could seemingly be met without lowering the American standard of living, but only with substantial economic growth. Robert Collins designated 1950 as a major demarcation in the American politics of economic growth: "By the end of 1950, the growth orientation that had been developed gradually since 1946 and articulated clearly in 1949 was firmly embedded in national policy."[15]

All these pro-growth politics and policies were not without effect, and postwar American economic growth is now a legend in the annals of capitalism. GNP increased from $282 billion in 1947 to $440 billion in 1960, as personal consumption expenditures went from $196 billion to $298 billion (all in 1954 dollars). It was during this time that the United States developed its distinctive reputation as a consumer culture, with families turning up in droves to purchase cars, appliances and television sets.

Internationally, there was a simultaneous phenomenon that brought economic growth to the fore of global politics, namely, the increasing agreement among the "developed" nations that there existed a "Third World" of developing nations that needed (whether their inhabitants knew it or not) economic growth. Many skeptics have since wondered if this recognition has been a good thing for the Third World, which they are wont to call the "Two-Thirds World." Some suspect the "First World" based its conclusions not so much upon Third World needs as felt therein, but more upon the ideals and mores of the First World judges. A more damning critique has been that the development programs financed by the World Bank and IMF have often been geared more toward increasing market outlets for First World corporations than toward improving conditions in the Third World.[16] This latter critique, for what it is worth, was behind the 1999 "Battle in Seattle" and subsequent demonstrations against the World Bank and IMF in Philadelphia and Washington, DC. Yet it is hard to sit in judgment of those international observers in the 1950s who saw living conditions that they thought, often correctly, were depressed, unhealthy and demeaning.

The international concern with economic growth and for the Third World had a parallel development in the United States. The

CEA found a microcosm of Third World poverty in certain sections of rural America. The goal was to alleviate poverty in these areas and contribute to national economic growth at the same time.

No sooner had economic growth become national policy, however, than it became highly contested. Dwight D. Eisenhower was elected president in 1952. He, a Republican Congress and the CEA put their emphasis not on growth but stability. This emphasis was not motivated out of environmental concerns. They were acting out of a "visceral fear of inflation and a keen sensitivity to the political dangers of recession."[17] It was understood then, as it is now, that an overly ambitious policy of growth was likely to produce a recession at the end of a spurt of growth.

But there was no explicit fear of any *limits* to growth. In fact, tending to inflation and stability first was simply a means of facilitating further, steadier growth. As Eisenhower put it: "I believe that economic growth in the long run cannot be soundly brought about except with [price] stability."[18]

The Republican regime of the 1950s was largely successful, or perhaps lucky, in keeping inflation reasonably under check. This success apparently came at the expense of growth rates, however. GNP grew at an annual rate of only 3.5 percent from 1954 to 1960, a rate today's politicians would brag about but noticeably lower than the 3.8 percent growth rates from 1947 to 1954. Economic growth, therefore, became a primary political issue during the 1960 presidential campaign. The Democratic Party adopted a goal of 5 percent annual economic growth in its platform.

While it may surprise Americans to hear of its "big business" party being beat to the punch in emphasizing economic growth, Republicans during the postwar years were developing their distinctive critique of government spending. The New Deal and World War II had been nails in the coffin of laissez faire and government showed no signs of declining. Federal spending averaged 17 percent of GNP from 1947 to 1960; the highest level it had reached during the New Deal was 11 percent.[19] Republicans began to blame rampant government spending for inflation.

In defense of their emphasis on economic growth, the Democrats argued that growth was required to maintain high levels of employment. Eisenhower countered that employment was important, yet not as important as inflation, which affected the material well-being of *every* consumer, employed or not. Kennedy won the 1960 election, and the Republicans learned a lesson. From then on, they would compete with the Democrats in the promotion of economic growth, and they would learn to combine promoting growth with objecting to big government. By the 1970s, they would earn their identity as the pro-industry, pro-wealthy and aggressively pro-growth party.

Meanwhile, the economic nature of the Cold War was becoming obvious, especially given the rhetoric of Soviet Premier Nikita Khrushchev, who pronounced: "Growth of industrial and agricultural production is the battering ram with which we shall smash the capitalist system."[20] Such rhetoric was designed largely to attract the Third World to communism, and was partly successful. Furthermore, the rhetoric was not entirely unfounded. Although comparisons were difficult for many reasons, the head of the CIA testified in 1959 that Soviet GNP had been growing twice as fast as American GNP throughout the 1950s. These were victorious times for the Soviets, who sent the first man into outer space in 1961.

No later than this point in American history, non-governmental organizations (NGOs) entered the macroeconomic political arena. The Rockefeller Brothers Fund produced a report in 1958 which portrayed economic growth as the way to achieve whatever American society might aspire to and called for five percent annual growth. This set the stage for future growthmania among the likes of the Competitive Enterprise Institute, American Enterprise Institute and National Manufacturers Association.

In terms of economic politics and policy, the 1960s would echo the New Deal of the 1930s. There were two major differences, however. First and from the outset, economic growth was now firmly entrenched as national policy, having constituted a major plank in Kennedy's prevailing platform. Second, the Vietnam War would sabotage Kennedy's Keynesian scheme. While World War II had taken the American economy to new heights, clearing markets and increasing capacity, Vietnam would simply be a drain on American wealth, health and morale, including consumer confidence. This was an early indicator of a full-world economy.

In the interim, however, Kennedy, and then Lyndon Baines Johnson, presided over a period of unprecedented government involvement in the economy and domestic affairs in general. Kennedy's program was called the "New Frontier;" Johnson's was called the "Great Society."

In 1964 James Tobin, one of Kennedy's economic advisors, observed that, "in recent years economic growth has come to occupy an exalted position in the hierarchy of goals of government policy."[21] Kennedy created a Cabinet Committee on Growth. At the Department of Commerce, signs in the halls interrogated the marching bureaucrats, "What have you done for growth today?"[22] Kennedy's New Frontier policies included a massive tax cut (to stimulate

FIGURE 9.1. The problem of uneconomic growth in the United States started to come into focus in the 1960s. John F. Kennedy (*left*) questioned his appointees and bureaucrats with, "What have you done for growth today?" By 1968, Robert F. Kennedy (*right*) warned against using GDP as a metric of social success. Credits: (*left*) Abbie Rowe, National Park Service; (*right*) U.S. News and World Report

personal consumption), a tax credit for capital outlays and a lowering of long-term interest rates to encourage capital investment. Johnson's Great Society was more like the New Deal. Originating as the War on Poverty, it firmly attached the label of liberal to the Democratic Party, because in the Great Society the government spent liberally on a liberal collection of causes. Education, health care, civil rights, urban renewal and some of the first ventures into environmental protection—all reached unprecedented levels in the federal budget. Social welfare expenditures amounted to nine percent of GNP by 1964.[23]

Yet the Great Society should not be viewed in retrospect as a "tax-and-spend" charity program. While it began as the War on Poverty, the war was about more than poverty. As Robert Collins put it, "The goal of the War on Poverty was not simply to enrich the poor but rather to change them so that they, too, could then contribute to the national goal of increased growth."[24] At the same time, however, the Democratic Party was coming under the influence of the historian Arthur M. Schlesinger and economist John Kenneth Galbraith. They questioned the further merits of increasing economic quantity, emphasizing instead the issue of socioeconomic quality. What did increasing GNP signify for American society? Was it still a healthy goal, conducive to well-being and happiness? Or had the American economy matured beyond such a simple metric of success, needing now a more sophisticated analysis? Schlesinger and Galbraith wrote explicitly on these issues, especially addressing the Democrats, who listened and indeed adopted a more sophisticated approach to the assessment of economic welfare. While the Republican Party was taking over the priesthood of economic growth, the Democrats were embarking upon a new holistic journey of economic welfare. This was a defining period in American party politics and went a long way toward establishing the current economic identity of the two American parties.

One of the new aspects of economic welfare considered by the Democrats was environmental protection. Three key environmental issues arose during the 1960s: air pollution, water pollution and endangered species. At this point in history, prior to the obfuscations of "human capital," "endogenous growth theory," and the "information economy," the connection of economic growth to pollution, endangered species and other high-profile environmental problems was readily apparent. Countering the big-business NGO community, newly formed environmental NGOs began to remonstrate against economic growth. Friends of the Earth, for example, warned of "the runaway US growth economy" and called for a "thorough reassessment and reversal of unlimited economic growth as a national goal."[25] Congress, too, had a clear understanding of the trade-offs. In the first sentence of the Endangered Species Act of 1973 Congress declared that "various species of fish, wildlife, and plants in the United States have been rendered extinct as a consequence of economic growth" (with the implicit caveat of technological progress as described in Chapter 8). Suddenly, it became fashionable to question the goal of economic growth.

Yet economic growth had come to be viewed as a panacea, and panaceas are not easily abandoned. While common sense indicated that economic growth caused environmental problems, a narrower, short-term thread of common sense also suggested that economic growth could provide the "resources" to attack any and all problems. This was the thread of truth in the fallacious fabric that came to be called the environmental Kuznets curve (Chapter 8). As Robert Collins described it: "The desire to use economic growth to transcend economic growth was as noble as it was chimerical, and the attention to growth's environmental consequences was as responsible as it was ironic. Still the driving optimism remained: Growth would make the chimerical and the ironic possible. On the horizon, however, lay a confrontation with national mortality, with limits, with Vietnam."[26]

Vietnam constitutes another pivotal episode in the politics of economic growth. The US had nearly two decades of unparalleled economic growth under its belt and remained supremely confident since the decisive victory in World War II. But with the US in the midst of the Great Society, its budget stretched, the North Vietnamese unleashed the Tet Offensive. Any doubt about the level of American involvement in Vietnam was dispelled. The United States took on a massive military—and therefore fiscal—responsibility. It was too much for the American economy to handle.

Under tremendous pressure and with crushing disappointment, President Johnson was forced to raise taxes, slash spending on the Great Society and ride the fence on Vietnam, having neither the resources to fund a clear victory nor the political option of an early withdrawal—three policies he had most disdained. Worst of all, inflation and recession rocked the economy at the same time. This was not supposed to happen. For Keynesians, especially, the conventional wisdom was that inflation was necessarily associated with an overheated economy, not with a downturn.

The unprecedented combination of inflation and recession generated a new term: stagflation. Stagflation caused consumers, investors and policy makers to panic, which only made matters worse. Furthermore, the value of the dollar in international currency markets was declining, contributing to a rush on gold reserves that threatened the stability of currencies and economies worldwide. The complexities of these international monetary developments are far beyond our purposes, but I note the confluence of inflationary, recessionary and monetary pressures to drive home the point that 1968 marked the most daunting economic crisis since the Great Depression. The relevance here, in a book primarily about sustainability, should be clear: no sooner had the American environmental movement appeared than the economy came back to dominate domestic politics. Notions of an environmental Kuznets curve turned out to be as politically naive as they were technically fallacious.

All this at the dawn of the presidency of Richard M. Nixon, one of the most clever, influential and darkly mysterious of American presidents. Nixon was nothing if not a masterful politician, and he was a prototypical advocate of what came to be called "smart growth." Fully recognizing the conflicting politics of economic growth and environmental protection, he expressed the need "not to abandon growth but to redirect it."[27] This was the classic win-win rhetoric that would come to dominate economic growth politics, Republican and Democrat, by the 1990s.

In terms of policy, however, Nixon moved further toward growth throughout his presidency. At first inflation was the primary concern and Nixon (who had served as Vice President under Eisenhower) oversaw a policy of gradualism, the intent of which was to gently slow the economy's growth phases to curb inflation. Unlike Eisenhower, however, Nixon thought unemployment was an even greater problem than inflation. Inflation had an insidious effect on everyone but often resulted in no clear reaction by voters. The unemployed, on the other hand, would invariably take their problems to the voting booth.

In 1971 Nixon established his New Economic Policy, prioritizing economic growth. Fiscal and monetary tools were brought into service. Nixon championed a full-employment budget to put the nation in a spending mood, and coerced the Federal Reserve into aggressively increasing the money supply. Equally aggressive tax reforms were designed to stimulate investment. Nixon established goals in terms of GNP.

However, inflation would simply not go away. While Nixon spent liberally to get people employed, he also instituted wage and price controls to control inflation. It was an extremely confusing episode in the history of American economic policy. Furthermore, and related to inflation, Nixon was faced with the additional concern of a declining dollar in the world's currency markets. This again put pressure on the gold reserves because dollars were still convertible to gold. Despite the potential for losing face in the Cold War, Nixon declared an end to the gold standard, cutting out the core of the Bretton Woods agreement. As we know, however, the World Bank and IMF lived on and evolved.

The fact that Nixon recognized limits to growth, or at least limits to growth rates, was reflected in his international policy of détente. Détente, meaning a relaxation of tensions, was presented as sophisticated diplomacy in the service of peace. It must have been clear to the superpowers, however, and to the rising powers in Europe and Japan, that little had changed in the fundamental relationship between economic growth and cold warfare. By reducing American military involvement in the world theatre, especially in Vietnam, the United States was able to retrench its economic forces and re-nourish the horse that pulled the Cold War wagon.

One more facet of Nixon's economic leadership is particularly relevant. Nixon was dramatic, a big-picture thinker with grandiose designs — in retrospect, some would say "delusions." He was highly intelligent and grasped the intricate linkages among social, political, economic, environmental and even spiritual affairs. The legacy he hoped to leave was "the lift of a driving dream," a new purpose to supersede the soul-searching of the Vietnam era. Consistent with détente, the driving dream was peaceable but vigorous. It called for economic ambition, competition and production to unite Americans and propel them to Cold War victory. With Nixon invoking a Protestant work ethic, economic growth became more than patriotic; it acquired a pseudo-religious verve. Unfortunately for Nixon, his ambition of shaping the national spirit sank in the wake of the Watergate scandal. Faced with impeachment, he resigned in 1974.

Ironically, one of the most sincerely spiritual of American presidents would soon take office and be profoundly critical of economic growth as a national goal. Following the uneventful completion of Nixon's vacated term by Gerald Ford, President Jimmy Carter exhibited the humility the United States needed on the heels of Nixon's nihilistic threat to democracy. Carter's humility, along with his agricultural upbringing, helped him to recognize the social and ecological costs of economic growth more than any president before or since. Furthermore, Carter had read E. F. Schumacher's *Small is Beautiful*,[28] was familiar with the Club of Rome's *Limits to Growth*,[29] and in his inaugural address stated, "We have learned that 'more' is not necessarily 'better,' that even our great nation has its recognized limits, and that we can neither answer all questions nor solve all problems. We cannot afford to do everything."[30] Carter was the president least likely to succumb to the self-sufficient services fallacy, the environmental Kuznets curve or other neoclassical notions of perpetual growth.

Nothing illustrates Carter's recognition of the conflict between economic growth and environmental protection better than his land conservation efforts. Carter rivaled Theodore Roosevelt in the lands he took out of commercial production and preserved for their ecological, aesthetic and recreational values. In one sense, Carter went further, enabled to do so by the Wilderness Act of 1964. Roosevelt had set aside national forests, parks and wildlife refuges. Carter did likewise, but his biggest contribution was his support for the Alaska National Interest Lands Conservation Act, which he signed on December 2, 1980. The act set aside nearly 104 million acres of national parks, monuments, wild and scenic rivers, wilderness areas and wildlife refuges. Much of this conservation land was designated wilderness pursuant to the Wilderness Act, giving it the strongest protection from economic growth under any American law (with the possible exception of critical habitat designated pursuant to the Endangered Species Act). Carter was also a staunch supporter of clean air, clean water and endangered species protection.

Unfortunately for Carter, us, and the grandkids, the economy of the 1970s did not provide a politically viable starting point for steady statesmanship. If Carter had come into office during a time of full employment and low rates of inflation, he probably would have done much to engender an American conservation ethic. Such stars were not aligned. Stagflation ran a ten-year course beginning in 1973, and by the end of Carter's term, Americans were again susceptible to growthmanship. They found it in Ronald Reagan.

Reagan, the movie star turned politician, was immensely popular. Americans had had enough of Carter's humility and welcomed Reagan's unbridled optimism and brazen persona. Like the macho character he often portrayed in the movies, he stormed into the international theatre by calling the Soviet Union the Evil Empire, a phrase he would repeat often and unabashedly.

Reagan's economic policies were fresh enough to warrant a name, Reaganomics, though they were known more generally as supply side economics. Supply side economics emphasizes the importance of facilitating production for the sake of economic growth. The Keynesian emphasis on using the federal budget to stimulate consumption had run its course, for the time being. Not that Reagan discouraged personal consumption—quite the contrary—but his way of engendering consumer confidence was to stimulate rampant production. Real estate developers are fond of saying, "Build it and they will come." The supply-sider says, more generally, "Produce it and they will consume."

Reaganomics was more than just supply side, however. It was a philosophy, an ideology that praised the market while deprecating big government. Reagan promised to "get the government off our backs," and by "our backs" he especially meant the backs of big business. He presided over the largest tax cut in American history, a "permanent" cut that was heavily regressive, meaning it favored the wealthy, especially wealthy investors. It also favored corporations: Reagan did whatever he could to free big business from regulations designed to protect the public. Environmental regulations, especially, were rolled back or loosely enforced. Once government was off the backs of big business, big business proceeded onto the backs of the rest of Americans, including little businessmen, environmentalists and, as many of us recall, anyone who tried to book a flight with a commercial airline.[31] Many little businessmen didn't seem to mind, however. For one thing, they too were opposed to environmental regulations. Furthermore, they invariably aspired to be big businessmen.

While these pages cannot amount to more than a cursory sketch of Reaganomics, many Americans and economists worldwide would shudder if I failed to mention a third and ironic feat—massive military spending that contributed to the biggest deficits in American history (at least until very recently). The national debt tripled during Reagan's presidency[32] and would hover over American politics for a decade, making it exceedingly difficult for the public or polity to consider any economic policy other than growth until the debt could be covered.

To be fair, one must acknowledge that Reagan's rampant militarization was instrumental in breaking the back of the Soviet warhorse. For our purposes, the significance is profound, especially in the context of a Chinese government becoming less bellicose and more businesslike. The chariot race was over at last! Cold War victory could have paved the way for a golden opportunity in the 1990s; the United States could have pursued a nobler purpose. Sadly, the opportunity was wasted upon a public forever pelted by the mass marketing of Wall Street and by a subsequent president whose major contribution to growth politics was the claim: "There is no conflict between economic growth and environmental protection."

It may surprise younger readers or older Americans with short memories that this classic win-win rhetoric was not the hallmark of President George H. W. Bush, Reagan's Republican successor. (We may skip the Bush presidency, which offered nothing original to economic policy or politics.) It was, rather, the calling card of President William J. Clinton, a Democrat. For eight years, Clinton curried favor among environmentalists by teaming with Al Gore. Gore, among his many distinctive accomplishments, was the author of *Earth in the Balance*. It is nearly impossible to read *Earth in the Balance* without concluding that Gore, perhaps more acutely and intelligently than any major politician in world history, understood the conflict between economic growth and environmental protection.[33] For an ecologist and ecological economist like me, then, it was a particularly peculiar episode in American politics when, right in the shadow of the Washington Monument, Gore echoed Clinton's rhetoric: "There is no conflict between economic growth and environmental protection…" This happened to be on Earth Day, 2000, six months before Gore's improbable loss to George W. Bush (son of the first President Bush). A general pall descended over the crowd, and thenceforth the assembled thousands became much cooler toward Gore and much warmer toward Ralph Nader and his running mate, Winona LaDuke, the Chippewa Indian woman who later belted out a courageous speech so laced with truth as to justify the event's proximity to the Lincoln Memorial. Earth Day 2000 featured the contrast between conventional win-win rhetoric and the Nader camp's penchant for telling it like it was. Gore's environmental leadership was tarnished if only a bit, but perhaps enough to change the course of history.[34]

To be fair to Al Gore, no one would take a braver stance on global warming, most notably exhibited in the documentary *An Inconvenient Truth*. In my opinion there is little question that the environment, the nation, and the world would be a better place today if he had been elected in 2000. I voted for Nader, but it was an easy symbolic gesture, with Virginia (my state of residence) solidly Republican at the time. (While punching a chad for Nader, I secretly hoped Florida would elect Gore.) Yet the fact remains, at the time of this writing in 2013, that the biggest environmental truth is still too inconvenient for any prominent Democrat to state. In *An Inconvenient Truth*, Gore came close, pointing out the perils of over-consumption. Yet he still felt compelled to suggest that economic growth could be reconciled with a healthy planet. The rhetoric was much softer, however, and perhaps there is still hope for steady statesmanship from Gore. It won't be easy, because he went on record with the win-win rhetoric, and revoking a declaration is not something that comes easy for a politician.

Having spent most of my adult life in public service, always with a conservation agency, I can testify that many professional conservationists look back fondly at the Clinton/Gore White House. Clinton did have numerous positive environmental impacts. He established the Grand Staircase-Escalante National Monument in Utah, thus keeping nearly two million acres out of the "supply side," at least for the time being. He inherited a hot potato in the Pacific Northwest, where the northern spotted owl had been protected pursuant to the Endangered Species Act. Under tremendous pressure from the timber industry, he generally upheld the law. His appointments to key cabinet positions were strong environmentalists. Nevertheless, the reality is that Clinton did nothing to foster our understanding of the perils of uneconomic growth. In fact, he did quite the opposite, seducing the nation into the folly of thinking it could reconcile economic growth and environmental protection. Insult was added to injury when some in the Clinton Administration ratcheted up the rhetoric with statements like: "Some people just don't get it! There is no conflict…" We will probably never know if Clinton himself actually believed this malarkey. It is possible he was swayed by his neoclassical advisors. It is also possible he kept a straight face by toying with semantics in his mind. ("It depends what you mean by 'growth.'") Perhaps it was simply a fib; sadly, Clinton was found lacking some scruples in that regard. In any case, his unrepentant win-win rhetoric about economic growth and the environment undid the good he did for the environment and posterity's economy. It helped turn the American public into a poster child for conspicuous consumption, setting an example that we now see manifesting in other parts of the world.

Of course, Clinton wasn't the first or last Democrat to serve the corporate community with his rhetoric. In the American campaign financing system, wealthy contributors play such a heavy role that virtually all politicians are beholden to Big Money. This sets up an "iron triangle" around the economic policy arena.[35] The concept of the iron triangle is standard fare in political science and refers to a special interest, a political faction endeared to the interest and a profession (usually manifested in a government agency) "captured" by the political and economic factions. One of the first iron triangles to be exposed was the military-industrial complex of American weapons manufacturers, congressional representatives from the manufacturing states and the US Department of Defense. President Eisenhower dramatically and courageously exhorted the United States in 1961 to beware this dangerous development.[36] Throughout the political economies of the world, iron triangles are a continual challenge to democratic governance, surrounding policy arenas and fending off all comers. Iron triangles are not necessarily conspiratorial, emerging organically out of mutual self-interests, but they inevitably result in the corruption of policy.

In the United States, the macroeconomic policy arena is surrounded by the biggest and most insidious iron triangle on Earth. The special interest is essentially the entire corporate community. Corporations are served, especially in the short term, by a theory of perpetual economic growth. The political faction includes virtually the entire political community. Big money gets politicians their jobs, and corporations have the biggest money. To complete the iron triangle we have neoclassical economists, including the influential Council of Economic Advisors, the powerful Federal Reserve System and the well-endowed Department of Commerce. Where do these economists come from? Straight out of academia, where their research was funded primarily by Big Money. We saw in Chapter 4 how land-baron wealth corrupted neoclassical economics from the outset. Would it make sense to think that the influence of money on the economics profession has disappeared or even diminished? No, the economists will continue touting theories of perpetual growth, the politicians will continue pumping their win-win rhetoric and, as long as the iron triangle fends off all comers, the corporations (and banks) will continue to accumulate wealth at the expense of posterity. That is why it is so important for ecological economics to take the place of neoclassical economics, especially in macroeconomic policy matters, with economic growth at the crossroads. It's the best hope for breaking through the iron triangle.

The iron triangle of economic growth flourished under the two terms of President George W. Bush. His knowledge of macroeconomics seemed primitive at best, and he was completely dependent upon his economic advisors for economic policy development. Meanwhile his rhetoric on economic growth was inherited from his father, and he was as pro-growth as any president in American history. Despite his "hands-across-the-aisle" campaign pledge, his obsession with economic growth led him into constant confrontation with the environmental community. He opened additional public lands to logging and mining, weakened air and water pollution efforts and supported a general gutting of the Endangered Species Act. (The act wasn't formally gutted—the Senate didn't go along—but it was effectively gutted through weak implementation and enforcement.) His reluctance to acknowledge the very existence of global warming, much less its relationship to human economic activities, was a diplomatic embarrassment. His reaction to international diplomacy toward reducing greenhouse gas emissions was summarized by stating: "The American way of life is not up for negotiation," echoing his father's words leading up to the 1992 environmental summit in Rio de Janeiro. His eagerness to wage war on Iraq was transparently related to a desire for cheap oil to keep the American economy growing, especially prior to his reelection campaign. The list goes on and on, as I well know, having served in the American bureaucracy throughout his two terms. I tried to develop some public education initiatives about the trade-off between economic growth and environmental protection, but I was severely suppressed as a result. The collective response from the hierarchy was, effectively, "Not with this president you don't."

When Barack Obama was elected President in 2008, morale in the conservation agencies immediately improved. I'd venture to guess that my morale improved more than most, based largely upon an article in the *New York Times Magazine* called "Obamanomics."[37] The author was David Leonhardt, an economics columnist for the *Times*, who reported a conversation with Obama on his campaign plane:

> "Two things," he [Obama] said, as we were standing outside the first-class bathroom. "One, just because I think it really captures where I was going with the whole issue of balancing market sensibilities with moral sentiment. One of my favorite quotes is—you know that famous Robert F. Kennedy quote about the measure of our G.D.P.?"
>
> "I didn't, I said."
>
> "Well, I'll send it to you, because it's one of the most beautiful of his speeches," Obama said.
>
> In it, Kennedy argues that a country's health can't be measured simply by its economic output. That output, he

said, "counts special locks for our doors and the jails for those who break them" but not "the health of our children, the quality of their education or the joy of their play."

The second point Obama wanted to make was about sustainability. The current concerns about the state of the planet, he said, required something of a paradigm shift for economics. If we don't make serious changes soon, probably in the next 10 or 15 years, we may find that it's too late.

Paradigm shift—you bet! And given that the explicit context was sustainability and the state of the planet, surely the paradigm shift would entail debunking the myth of perpetual economic growth. As Leonhardt subsequently ruminated:

> Both of these points, I realized later, were close cousins of two of the weaker arguments that liberals have made in recent decades. Liberals have at times dismissed the enormous benefits that come with prosperity. And for decades some liberals have been wrongly predicting that economic growth was sure to leave the world without enough food or enough oil or enough something. Obama acknowledged as much, saying that technology had thus far always overcome any concerns about sustainability and that Kennedy's notion had to be tempered with an appreciation of prosperity.
>
> What's new about the current moment, however, is that both of these arguments are actually starting to look relevant. Based on the collective wisdom of scientists, global warming really does seem to be different from any previous environmental crisis. For the first time on record, meanwhile, economic growth has not translated into better living standards for most Americans. These are two enormous challenges that are part of the legacy of the Reagan Age. They will be waiting for the next president, whether he is Obama or McCain, and they'll probably be around for another couple of presidents too.

Leonhardt was on the right track, but the enormous challenges will be waiting for more than another couple of presidents. With economic growth at the crossroads, no president from here into the foreseeable future would be serving the country by pulling out the stops for economic growth. Rather, they would be contributing to uneconomic growth, causing more problems than they solved, threatening posterity more than securing its future.

The other shortcoming of Leonhardt's analysis is equating a steady state paradigm with a "liberal" agenda. If only the liberals deserved such credit! As we've seen in this chapter, liberals and conservatives, Democrats and Republicans, communists and capitalists—all have been hell-bent on growth at all costs.

As a longtime conservationist and current political independent, I especially bristle at the pilfering of the word *conserve* by so-called "conservatives" who want to double the rate of growth and double the size of the American economy. According to the Merriam-Webster Dictionary, "conserve" means "to keep in a safe or sound state…especially: to avoid wasteful or destructive use." The first example given? To "conserve natural resources." Obviously! "Conserve" goes with natural resources like "defend" goes with country and "safeguard" goes with Constitution. That's why those who conserve natural resources are called "conservationists," and the political manifestation of the conservationist is what would warrant the label "conservative." In other words, anything politically conservative ought to refer, *especially*, to the conservation of natural resources. Certainly, the word "conservative" should never be used to refer to the *non*-conservation of natural resources.

Therefore it really takes the rhetorical cake that the word "conservative" has come to mean such an anti-environmental, pro-growth, transform-the-world-into-plastic agenda. It always seems to be self-proclaimed "conservatives" that want to roll back environmental protections. "Conservatives" push for drilling in the Arctic. "Conservatives" want to gut the Endangered Species Act. "Conservatives" don't want to limit greenhouse gas emissions. Hummer drivers and Yukon drivers like Glenn Beck call themselves "conservatives."

"Conservatives" my keister!

I can't rectify decades of rhetorical sabotage in one paragraph, but let me just clarify what it really means to be conservative, or at least what it's really supposed to mean. In a world of plain-spoken truth, I am conservative and my friends are conservative. We're the ones who conserve natural resources. We ride our bikes to work, sometimes drive the hybrid (or some smallish car) and always shut the lights off when leaving home or office. We vote for politicians who are strong on conserving natural resources. We volunteer for organizations that help protect the environment, which amounts to conserving natural resources. And we don't appreciate the gas hogs stealing our identity, calling themselves conservatives and hiding a sow's ear in a silk purse.

In any event, what has become of Obamanomics? As I write in January 2013, at the dawn of Obama's second term, theoretically there should be more room within the US government for open discussion about limits to growth and the conflict between economic growth and environmental protection. However, the bureaucracy is still top heavy with old win-winners on one hand and growth-at-all-costers on the other. Legitimate, science-based efforts to raise public awareness of the trade-off between economic growth and environmental protection are still suppressed and even penalized. Typically suppression starts out with a verbal "gag order" that prevents an employee from working on a topic or even talking about it. If the employee doesn't take the hint, or take it seriously enough, the gag order may be put into writing and euphemistically called a "memorandum of expectations," which may be in effect for years. If the employee feels strongly enough about the topic to persist, the gag order is grounds for a formal reprimand, which then becomes grounds for a suspension. It's like a sword of Damocles that dangles closer and closer to the scalp. Soon the employee is right on the cusp where an untied shoe could be grounds for termination.

There are other ways besides career-threatening disciplinary action to prevent a civil servant from talking about limits to growth or the conflict between economic growth and environmental protection. His or her job title may be changed to something sounding less relevant to economic growth. The job description may be modified to allow only for dealing with specific, narrow topics. There's also the proverbial—and literal—office without windows, among other things.

That's how sound science is suppressed in 21st-century American government: blunt suppression and threats of termination coupled with morale-sapping humiliation. Yet the true civil servant persists, because civil service remains a crucial crucible for advancing inconvenient truths. It's not the only crucible, but it is a crucial one. If federal agencies with missions such as wildlife conservation, environmental protection and sustainable development cannot develop the wherewithal to deal openly and explicitly with the challenge of economic growth, what hope do we have for the rest of the polity, dominated as it is by neoclassical economics and perpetual growth theory?

Meanwhile Obama himself has yet to provide any clear steady statesmanship. During his first two years in office, his rhetoric on growth and the environment was neo-Clintonian, although not as brazenly win-win. It was more along the lines of "greening the economy," and the President steered clear of blatantly fallacious win-win rhetoric. His economic and environmental agendas had clear and separate goals. His economic focus was on rescuing the financial system and creating jobs; he seldom used the phrase "economic growth." Meanwhile, he promised to protect the environment, period, and the BP oil spill gave him a platform (pardon the ugliness of the pun) to put the environment first.

Some would argue that Obama was necessarily promoting economic growth when he bailed out the banks in 2009 and called for job creation. But they wouldn't necessarily be right. Bailing out banks and saving the insurance industry was necessary for stabilizing the financial system, which needed to happen with or without economic growth. It was needed especially to protect the modest lives of relatively innocent borrowers and bank customers (even though wealthier swindlers may have benefited the most).

As for jobs, it is true that GDP growth is seen as a job creator through the lens of conventional economics. One of Obama's noisier economic advisors, Lawrence Summers, certainly helped to circulate that message in the media. Technically, though, more jobs can be created while capital expenditures decline (to be further explored in Chapter 11), so employment can increase without growing GDP. Therefore, a president can call for more jobs without necessarily promoting economic growth. Such "labor intensification" has its limits, naturally enough, but it can solve short-term unemployment problems while more important issues are dealt with.

And what issues would be more important than full employment? For starters, how about full employment for your kids, say five years from now, or for your grandkids in a couple of decades? How about the environment—air, water, soil, minerals, timber, fisheries, etc.—the foundation and building blocks of the economy? How about the other species on the planet?

Unfortunately, it's too easy for critics to hone straight in on "other species" and rant, "Who cares about other species—we're talking about the economy!" But we better care, because these other species are like canaries in the coalmine of the grandkids' economy, and we've been shooting them down like targets at a county fair. Splat goes the spotted owl, poof goes the polar bear, 1,372 federally listed species on the ropes and, with very rare exceptions, down for the count. It's sort of like your fellow bank customers going bankrupt all around you. How secure does that make you feel about the bank?

Yet because the paradigm shift Obama referred to has not yet happened, and we don't have a public aware of uneconomic growth, deficit spending and other desperate measures to "stimulate the economy" must seem like the only political option. On the other hand, Obama could be more forthcoming with the American people—and people worldwide—regarding the ultimate unsustainability of growth and the concept of uneconomic growth. He could be leading the paradigm shift, not waiting for it to happen.

So it must have been a sad day in the saddle when, on January 18, 2011, Obama rode out onto the tantalizing trail of the win-win slippery slope. (If it wasn't sad, it certainly was cynical.) In an op-ed for the *Wall Street Journal*, Obama promised that "federal agencies (will) ensure that regulations protect our safety, health and environment while promoting economic growth." In other words, we would have our cake (the environment) and eat it too (for economic growth), and federal agencies would be there to dish it all up! It was an inconvenient day for the truth, especially down in the crucible of civil service.

Fortunately, the days of win-win rhetoric are numbered. Recognition of the fundamental conflict between economic growth and environmental protection is gaining prominence in the science journals, bookstores, academic departments, NGOs and even some government agencies. It's getting too obvious to miss. And to help focus attention on it, thousands of citizens including top scientists and other prominent figures have signed a position circulated by the Center for the Advancement of the Steady State Economy stating that "there is a fundamental conflict between economic growth and environmental protection."[38] You can't get any clearer than that.

So what we need now is a president who will parlay this knowledge into public support for policy reform. The president can clarify once and for all that we can't have our cake and eat it too. Can you almost hear him, or her? "We need to balance our concerns about environmental protection with our concerns about full employment, and that doesn't square with growth everlasting. What we need is a healthy, steady state economy balanced with a healthy environment, not an overgrown economy and a shrunken environment."

How would a president and other policy makers help create a steady state economy? Given popular support, first off, policies designed to "grow the economy" would be discontinued. Next, steady state policy tools (such as resource capping) would be employed. There is no shortage of policy options, as we'll find in Chapter 11. But the horse has to come before the cart. The steady state economy has to be a goal with widespread public support before a suitable policy framework can be constructed. Presidential leadership is needed to generate such support. Then, with widespread public support, a steady state economy would be engendered from the "demand side," too, with temperance trumping conspicuous consumption. That's the subject for Chapter 10.

Meanwhile, Obama has the rest of his presidency and presumably a number of productive years of public service afterward. It's not too late for him to be the Truth Teller in Chief. He's tested the slippery slope of win-win rhetoric—gotten his foot muddied a bit—but he hasn't committed himself to a mudslide yet. The trade-off between economic growth and environmental protection is perhaps the most inconvenient of all truths to acknowledge, but it's better than a full slide down the slippery slope of green growth rhetoric. That could be a legacy breaker.

CHAPTER 10

Hummer Haters: The Steady State Revolution Revisited

> *Nothing could be more salutary at this stage than a little healthy contempt for a plethora of material blessings.*
> ALDO LEOPOLD

"Hummer Haters." That was the title of an October 8, 2006, segment on *ABC Nightly News*. In a climate of developing angst over oil supplies, wars in oil-producing nations and high prices at the pump, Americans were getting sick and tired of the Hummer. Some of these Hummer haters had purely parochial reasons for their hatred, most notably the hogging of valuable space by Hummers in traffic lanes and parking spots. But many had more sophisticated, holistic and patriotic rationale. This lumbering pile of iron on oversized wheels—usually in yelling-at-you yellow or bad-boy black—represented much that was wrong with America.

First was gas hogging. With Peak Oil more or less imminent and gas prices on the rise for *all* Americans, how unpatriotic could a consumer get, to select one of the most—and most obviously and unapologetically—gas-hogging automobiles ever to roll out of Detroit?

A closely related concern was global warming. When we know beyond a shadow of a doubt that personal vehicle emissions are one of the leading sources of the greenhouse gases threatening our nation and planet, who are these people who would so willingly and wantonly select the very emblem of those emissions?

And finally, what about our national security, especially in an age of anti-American terrorism? Clearly this hinges on a disciplined, unselfish, temperate American image—not that those would be bad traits to aspire to in any age. Yet the Hummer not only guzzles the gas that other nations need for their daily bread, it is the civilian version of that symbol of American bullying, the Humvee. It really doesn't matter if the American civilian believes that the US is actually a bully or a buddy to Iraq, Afghanistan, the Middle East, or the entire international community. What matters is that entire generations in so many of those nations believe *precisely* that. How could an American be so crass as to rub it in their noses by commandeering the roads with the most militaristically evocative personal vehicle on the planet? It's like adding diplomatic insult to economic injury.

All this readily explains the burgeoning of the Hummer haters.

The ABC feature also explored the social manifestation of Hummer hatred. Perhaps the key aspect is the focus on *Hummer* hatred, per se, and not Hummer *driver* hatred, at least within the confines of the US. The feature engaged Hummer haters and Hummer drivers in mutual exploration of their respective opinions. On the one side, the hatred of the Hummer was real, but the attitude toward the Hummer drivers could be paraphrased by a quizzical quip, "What the hell were they thinkin'?" On the other side, the attitude seemed to be, "I like big, I can afford big, and it's nobody's business if I buy big." But they weren't thinking big, and now Americans—even ABC—were making it their business.

As I contemplated the news feature, it occurred to me that Hummer hatred was the quintessential example of the "steady state revolution," the subject of *Shoveling Fuel for a Runaway Train*. After a critique of economic growth and an overview of ecological economics, *Shoveling Fuel* provided a blueprint for the social movement needed if the US was to establish a steady state economy. A strong phrase was used to describe the movement—"steady state revolution"—because of the magnitude and pace of change that were necessary to avoid a truly devastating train wreck at the limits to growth.

The magnitude and pace of change are what separates a revolution from evolution. In the case of the steady state revolution, however, there is one other distinctly revolutionary aspect: a 180-degree turn in how Americans view a previously acceptable, and even emulated, behavior. That behavior is conspicuous consumption.

Shoveling Fuel got off to a promising start when the University of California Press published it in 2000. University presses seldom score bestsellers, and this was a book that neither Wall Street nor the Competitive Enterprise Institute would be sponsoring in the bookstores. Nevertheless, on the heels of the Battle in Seattle and other demonstrations against the World Bank and IMF, the time was ripe and it wasn't long until a few thousand copies of *Shoveling Fuel* were circulating. Feedback was positive, strong and growing, and for a few hopeful months, it looked like the steady state revolution had a chance to spark some sustainable tinder.

But history was shoveling fuel in another direction. On September 11, 2001, Al-Qaeda terrorists attacked the United States. The World Trade Center, a scion of the system that could have been reformed from the grassroots, instead went down in a surreally horrifying ball of flames. All other agendas were dropped and if ever there was a time for Americans to unite behind their president, this was it.

In response, George W. Bush, the President of the United States of America, told us to go shopping and traveling to Disneyworld. Many of us were appalled and thought this was one of the most embarrassing, insensitive and cynical statements ever uttered by a politician, much less a standing president. But this was no time, nor would it be for at least a few years, to show disunity. The cabinet and Bush's advisors stood by the president and, in subsequent months, parroted the calls for shopping and traveling. Sympathetic commentators pointed out that the Bush Administration was using shopping and vacationing as examples of economic growth. As a way to show the "enemies of freedom" our indomitable will, we were to keep producing and consuming goods and services like never before. The World Trade Center might be down, but not our propensity to consume.

In the context of the previous chapter, it is easier to understand how the American public could accept the call to shop without batting an eye. People had been pummeled for decades with pro-growth political propaganda and massive advertising campaigns, all underwritten by Keynesians and neoclassical economists. Soon enough, we had a new class of "patriotic shoppers," sometimes even referred to (even more embarrassingly) as "militant shoppers."

It should be emphasized that this was a time to support not only our president, but each other. We were to put aside our differences of race, creed, gender, politics and opinion to develop and demonstrate our unity to the enemies of America. Some of our long-standing disagreements among us couldn't be held in check for long, but there was no room for new ones.

If someone had tried to write a futuristic script of the steady state revolution, they could hardly have imagined a more thorough and untimely snuffing of its spark. The steady state revolution was all about *less* shopping and *fewer* trips to Disneyworld; suddenly, though, shopping and Disneyland were patriotic obligations. And while the steady state revolution entailed recognizing a class structure of conspicuous consumers on one hand and more conscientious consumers on the other, and a certain productive tension between the two, post 9-11 America was about abolishing such internal divisions. If anything, the conspicuous consumers were now a breed of "patriots" to join.

But history's book is never finished, and now we have reached a chapter, literally and figuratively, called "Hummer Haters." It is time to revisit the steady state revolution.

All revolutions in political economy and socioeconomics have a class structure. Unfortunately, some turn bloody: think of the French Revolution and the various communist revolutions that have pockmarked the face of humanity. These were more like civil wars that often boiled over into actual "hot" or "cold" wars between nations. In thinking of the steady state revolution, let us put all such visions of conflict aside, except as reminders of what to avoid. There is no such wresting of power and capital in the steady state revolution. The classes in the steady state revolution are not based upon "ownership of the means of production," to put it in Marxist terms. The class structure of the steady state revolution is based upon one thing and one thing only: consumer behavior or, to be more specific, conspicuousness of consumption.

Perhaps the best example of a movement in the US that serves as a model for the steady state revolution is the anti-smoking campaign. By the time it was proven deadly, millions of individuals were addicted to tobacco and our culture was hooked on tobacco imagery: the rugged Marlboro Man, the chic chick with the Virginia Slims, the world-venturing Camel smoker. But once the hazards were documented, and even in the face of corrupt and collusive advertising by Big Tobacco, the common sense and virtue of citizens took over. Individuals quit by the droves, and a cultural stigma was attached to those who continued: they were ignorant, weak and self-destructive. And that was only the beginning. When second-hand smoke was found to be toxic to *others*, smokers acquired an additional reputation as selfish, uncaring and obnoxious. Smokers were socially castigated by non-smokers, so fervently that the castigation is now manifest in public policy. A growing list of public places are now smoke-free or allow smoking only in cordoned corners where smokers are peered at like animals in a zoo.

I can empathize with the smoker, having taken to tobacco as a teenager. I can also attest to the discomfort caused by the cultural turn against smoking, as the tobacco habit haunted me into young adulthood. But the social discomfort was a good thing, in retrospect. When you already want to quit for other reasons, such as the health of yourself, those around you, and those you might influence with your example, the likelihood of being castigated helps!

The parallels of the anti-smoking movement to the steady state revolution are uncanny: two social classes, with one shamed and castigated; a natural selection for non-smokers in everything from romance to careers; and, most importantly for our purposes, a drastic reduction in consumption. When we think about the steady state revolution, let us forget about coups and beheadings, and focus on the courageous, healthy and extremely effective example of the anti-smoking movement.

Of course, there are significant differences, too, between smoking and conspicuous consumption. For example, a person may readily be identified as a smoker. You either smoke or you don't, and when you're not smoking you tend to smell and have rotten teeth. In contrast, how do we know if a consumer is "conspicuous?" Almost all acts of consumption are readily observable, so "conspicuous" is a subjective and relative concept. Fortunately this is not a major problem because, by definition, *conspicuous* consumption is especially observable. It's reminiscent of the judicious observation, "I can't define pornography, but I know it when I see it."[1] The Hummer, the mansion on the hill, the floor-length fur coat—we all know truly conspicuous consumption when we see it.

On the other hand, many consumer goods and services defy quick and certain identification as conspicuous. Is driving a midsized SUV an act of conspicuous consumption? Wearing a rabbit-trim jacket? Getting a massage?

One way to sidestep this devil in the details is to hone straight in on the most conspicuous of conspicuous consumers. They might be called the "liquidating class" to signify the liquidation of natural capital entailed by their consumption, and let's say they comprise the upper one percentile in personal consumption expenditures. This designation has an important political advantage, because it is much easier to unify people—such as a large steady-state class of conscientious consumers—to oppose the behavior of a very small, exceptionally problematic faction than against a large group (say the top 20 percent in consumption) comprised, on average, of less problematic individuals. More on this in a moment, but first let's have a look at an example of liquidating behavior.

If this were the 1980s, the task would be simpler, for the television series *Lifestyles of the Rich and Famous* would provide one-stop shopping for examples of the liquidating class. All we'd have to do is convert those charming vignettes of the rich and famous into re-

Hummer Haters: The Steady State Revolution Revisited | 265

FIGURE 10.1. Hummers H3, H1 and H2 (*above, left to right*) and "McMansion" (*below*): symbols of conspicuous consumption and heavy ecological footprints in the USA. Credits: (*top*) Wikimedia Commons; (*bottom*) David Klotz

flections of monstrous ecological footprints. Today, however, there seems to be a more nuanced psychology involved in the display of conspicuous consumption. Many conspicuous consumers aren't so keen on advertising their wanton wastefulness to the general public. This probably stems from the fact that many of the most conspicuous consumers are bankers and corporate CEOs. In these days of financial crises that hobble the middle and lower classes (that is, the vast majority of folks), bailed-out bankers and CEOs don't necessarily want shareholders apprised of how the profits are getting spent. Besides, they aren't overly concerned with impressing the masses, intent instead on impressing their peers among the liquidating class. But word gets around—that's the nature of conspicuous consumption. And some of the liquidators are so ostentatious that they make little attempt to even feign any moderation. Dick Meyer of CBS called this "aggressive ostentation." [2]

Now just as I can empathize with the smoker, I can empathize with the liquidator. Most of us probably can to some degree. There are strong, even Darwinian, instincts for material display. Who among us has not showed off a bit with something, say some favorite clothing or a car, at some point? And of course, conspicuous or not, it is a rare individual indeed who has never taken more than the absolutely essential bite when another tasty morsel was well within grasp. So there is no room for feeling holier than thou when thinking about consumption. Besides, conspicuous consumption wasn't the threat to posterity that it's become in the age of Supply Shock, which crept into place rather unannounced.

That said, an attitude adjustment is clearly and desperately in order. We can no longer afford to ignore conspicuous consumption, much less emulate it. We have to analyze it, understand what it means for the kids and the grandkids, and respond to it appropriately in order to discourage such behavior. A hypothetical example won't do here, either. With all due empathy, we need a solid, real-life example to demonstrate the ecological impact and the psychological folly of liquidator-level consumption.

Our prize-winning example of an ostentatious liquidator — drum-roll please—is one Stephen A. Schwarzman, co-founder and CEO of the Blackstone Group, a private equity firm.[3] Schwarzman probably has numerous fine qualities, but conscientious consumption is not among them. Dubbed the "Golden Ass" by business columnist Daniel Gross,[4] he resides in an "apartment" at 740 Park Avenue in New York. In this case "apartment" needs quotation marks because most folks wouldn't think an apartment could cost 30 million dollars. But most apartments didn't previously belong to John D. Rockefeller, Jr., nor do they have the highest ceilings and widest hallways on Park Avenue, nor are the exteriors clad with limestone, etc.[5] Plus most apartment dwellers are tenants; Schwarzman owns the whole building. To be more accurate, though, Schwarzman resides there only part of the time, for he can also be found sunning by the pool at his 11,000-square-foot mansion in Palm Beach, Florida. Evidently he doesn't like to be confined; he's got other places too: "Lavish ones. In New York, the Hamptons [that's on Long Island, not the Park Avenue apartment], St. Tropez [French Riviera] and other posh places where wealthy people congregate."[6]

Schwarzman's executive chef in Palm Beach says he often spends $3,000 for a weekend of food for himself and his wife, including stone crabs costing $40 per claw. On weekdays he "expects lunches consisting of cold soup, a cold entrée such as lobster salad or fresh grilled tuna on salad, followed by dessert.... He eats the three-course meal within 15 minutes." This otherwise perfectly fine fellow (I assume) is hell on fisheries as well as housing materials, and that's only in his regular routine!

What punctuates the ecological footprints of the liquidating class is the gala affair. Schwarzman does a lot of punctuating, and one such affair was so extravagant as to make the *New York Times*. It was Schwarzman's 60th birthday party, held at the Seventh Regiment Armory on Park Avenue. The party was held just a few days after the Blackstone Group completed a $39 billion purchase of Equity Office Properties, the largest leveraged buyout ever (as of 2007 when the party was held). Among the 350 guests were Donald Trump (liquidator *par excellence*), John Thain (chief executive of the New York Stock Exchange Group), and Sir Howard Stringer (chairman of Sony). Even Paris Hilton was there, "surrounded by an admiring group of investment bankers from Bear Stearns, Lehman Brothers, and Goldman Sachs."[7] These folks wouldn't be settling for Ritz crackers, Steve's Cheese and a bottle of Schlitz. "The dinner included lobster, filet mignon, and baked Alaska, topped off with potables such as a 2004 Louis Jadot Chassagne-Montrachet."[8]

Nor would the guests be arriving by bike, bus or Volkswagen Beetle. "Out from a black Escalade stepped CNBC anchor Maria Bartiromo, with her husband, Jonathan Steinberg, son of financier Saul Steinberg..."[9] (For those who don't know American automobiles, the Escalade is the Hummer of the frou-frou set.) White stretch limos were the choice of those who didn't want to drive themselves.

Schwarzman's tromping through the trophic levels doesn't end with the agricultural, extractive, building or automotive sectors. He sucks up the services, too. At the birthday gala, "held in a hangar-like space that can accommodate thousands...festooned in red and white in a Valentine's Day theme, with orchids scattered everywhere," a catering service tended to "the entertainment and production staff."[10] Speaking of entertainment, no piped-in polka band would do for this one. The production was emceed by comedian Martin Short. Rod Stewart performed, apparently to the tune of a million bucks.[11] Composer Marvin Hamlisch did a number from *A Chorus Line*. Patti LaBelle and the Abyssinian Baptist Church Choir sang "Happy Birthday." The whole affair cost about $4 million.[12]

Many a professional journalist and amateur blogger have lampooned Schwarzman and his birthday bash, primarily from the angle that here was this financial guru, partying like it was 1999, more bullish than the bronze bull on Wall Street, not long before the financial crisis of 2008 commenced. The more conservative pundits also liked to point out that Schwarzman's showy spending raised the ire of legislators, who then went after the tax status of private equity firms, threatening not only Blackstone but the many wannabe Blackstones and *their* CEOs. Schwarzman got out relatively unscathed, however, and was eventually seen in the media as something of a Mr. Magoo, a smiley victim of circumstances.

Some journalists chose a different path and analyzed not the man's behavior but the institutional framework he lived in; not what made the man tick, but the clockwork he ticked to. "In other words," as Yvette Kantrow put it, "forget the crabs and the mansions and the birthday bash. What about private equity, which made Schwarzman's superluxe lifestyle possible? What are…readers supposed to make of that? Is it good for the economy or not? Does it create jobs or destroy them? Improve businesses or sap them? Are ordinary Americans better or worse off because of it?"[13] Those are good questions, too, but with economic growth at the crossroads, the last thing we can afford to do is "forget the crabs and the mansions and the birthday bash." Schwarzman is a surrogate for the liquidating class, the consumption of which needs to be thoroughly analyzed and pondered for its ecological and economic impacts, not vaguely ridiculed then swept under the rug of social malaise.

Surely a graduate student in sustainability studies can calculate the ecological footprint of the liquidator lifestyle (and I hope one does). How many planets would it take to support the human population if each one had a footprint such as Schwarzman's, for example? We can help set up the framework here. The average American's ecological footprint is approximately 9.7 hectares per person per year, while the global capacity is 1.8 hectares per person. In other words, if everyone in the world lived like an average American, we would need almost five and a half planets (9.7/1.8 = 5.38) to support us into the long-run.[14] Recall from Chapters 7 and 8 that we can temporarily live beyond our means by liquidating stocks of natural capital, such as petroleum. It's the kids and grandkids who will pay the ecological and economic debt.

But we're not all average Americans, are we, Mr. Schwarzman?

If we are already in debt by 5.3 planets to support the average American lifestyle, and Schwarzman consumes, say, 1,000 times as much as the average American, then it would take 5,300 planets to support the current planet's population of Schwarzmans! Shouldn't there be a law against that?

Someone coming out of the blue, straight to this chapter, might say, "That's not fair. It's not like Schwarzman is consuming *1,000 times* as many crabs as you or I." Well, I wouldn't be so sure. Given that $40-per-claw stone crab habit, and the fact that there are 365 days per year, Schwarzman could eat just a few crabs every few days and get up into the thousands quite quickly, while you or I might have crab once or twice a year. Our crabs would probably be more on the order of $4 a claw, too, and wouldn't require a personal chef for their preparation. But to get too concerned with the specifics of Schwarzman's crab consumption would be missing the bigger point: that is, what the trophic theory of money tells us about ecological footprints. That's why only someone coming out of the blue would be asking such a question at this point, because the rest of the readers would have learned about the trophic theory of money in Chapter 7.

One of the most important implications of the trophic theory of money is that GDP is a sound indicator of the ecological footprint. Global GDP is an indicator of the global ecological footprint, American GDP is an indicator of the American ecological footprint, and New York GDP is an indicator of the New York ecological footprint. Your personal GDP is an indicator of your ecological footprint. The quickest way to estimate your personal GDP—and thus your ecological footprint—is by estimating your consumption expenditures. The liquidator's outlandish consumption expenditures result in an outlandish ecological footprint. The liquidator is a bad citizen, in other words. He may have redeeming traits, like Mr. Magoo, but that doesn't make him a good citizen.

All of us waste a bit here and there, splurge a bit now and then, drive a car when we could easily walk, etc. But the degree to which we avoid waste and splurging is a measure of our citizenship, our ethics and our loving concern for everyone's kids and grandkids. With economic growth at the crossroads, those at the extreme end of the consumption spectrum are rapidly becoming socially unacceptable. The liquidator's lifestyle is getting to be a crime against the spirit of humanity. Laws *should* be passed to prevent such heavy stomping on the planet. Meanwhile, the only way to peaceably protect ourselves and posterity from the liquidating class is with social remonstration, denunciation and castigation. As Aldo Leopold said, "Nothing could be more salutary at this stage than a little healthy contempt for a plethora of material blessings."[15]

"Healthy contempt" sounds a bit oxymoronic, but Leopold was no moron and an ox needed goring. It's too bad for us that not only is the environment being degraded by liquidators, but that we have to experience the contempt aroused by their liquidating behavior. Contempt is a stressful emotion and no one enjoys it, neither from the receiving nor the giving end. Yet it is *healthy* contempt from the standpoint of protecting the environment and the grandkids. It is this contempt or castigation that will change the behavior of the liquidating class. Hopefully it is not the only thing that will change their behavior (for that will take a lot of contempt). They say you can catch more flies with honey than vinegar, so honey too is worth a try. Many liquidators might enjoy philosophical discussions of positive, sustainable visions. A percentage of them will follow up such discussions with action, and there are wealthy CEOs who became philanthropists while dramatically curbing their own consumption. However, others of the liquidating class wouldn't give a steady stater the time of day, much less modify their consumption behavior in response.

Contempt or castigation, on the other hand, is not so easy to ignore, especially when it catches on in the media or political circles. And if you doubt that castigation motivates peoples' behavior, consider that a core principle in psychology is that our behavior is steered by other people's opinions. Abraham Maslow (1908–1970) developed the theory that only our need for food and water, safety, and love and affection motivate behavior more than our need for self-esteem. And self-esteem is generated largely from the feelings of others; if everyone loathes you, you're highly unlikely to have much. Love and affection, too, clearly accrue as a function of how others view you. The needs for love, affection and self-esteem are extremely powerful motivating forces, and they prevent what would surely otherwise be a great deal more antisocial behavior. Laws provide a safety check, allowing us to incarcerate those who are psychologically sick enough to eschew love, affection and self-esteem for the sake of fulfilling carnal or violent urges. With economic growth at the crossroads, we must invoke the powers of Maslow's hierarchy of needs to make conspicuous consumption a behavior of a bygone era, something akin to clubbing a cavewoman or whipping a servant.

Some would probably object to this focus on the top percentile of consumers. They'd feel that tempering the consumption of this liquidating class would not amount to a drop in the bucket of sustainability. However, the objection misses two key points. First and most simply, the consumption of the liquidating class is no drop in the bucket (as Schwarzman so freakishly demonstrates). One transaction by one of these liquidators can trump (no pun intended) a steady stater's entire lifetime of frugality. Yet the second missed point is much more important. Let us assume that, consistent with

Maslow's hierarchy of needs, the castigation of the liquidating class will result in their adopting a different, less conspicuous and lower level of consumption. We come back to the objection that we are "only" reforming the behavior of one percent of consumers. That is indeed the case at the very beginning of the steady state revolution, but what happens when the liquidating class lowers its consumption? Well then it is no longer the liquidating class. Instead, we have a new liquidating class, comprised of consumers who had occupied the second percentile in personal consumption expenditures at the beginning of the steady state revolution. The castigation continues, perhaps with the participation of the ex-liquidators (whose social influence upon the new liquidating class will presumably be more pronounced). This "second-generation" liquidating class reforms its consumption behavior, the third percentile rises to the occasion (so to speak), and the cycle repeats itself indefinitely.

Of course this is only a model of reality, constructed for ease of explanation. In reality there is not such a herky-jerky replacement of classes, one percentile after another, but rather a movement at the margin, an erosion of the propensity to consume. Picture a pyramid of recently piled sand, settling and eroding. There is some movement of sand all along the upper surfaces, but especially at the top. We could say that it's due to the castigation of the liquidating class that the marginal propensity to consume decreases fastest at the top of the pile.

So the process continues indefinitely but not forever. It won't get to the point where each grain of sand is flush with the other. It won't even get to the point where everyone is living in mud huts and castigating the few who still have tin roofs. The extent to which the steady state revolution reduces consumption will be a matter of common sense, balance, priorities and political economy informed by studies in ecological economics that help identify a more optimal size of the economy and relate that to personal consumption expenditures. And hopefully it will be led by a sufficient combination of academics, citizens, civil servants, business leaders and politicians with the integrity and courage to endorse what is right for society more than what is profitable for corporations or expedient for political campaigns.

To summarize, the steady state revolution is not about placing the *entire* responsibility for American overconsumption on one percent of its citizens and forgetting about everyone else. Rather, the approach is to identify that one percent as the single *most* responsible faction for the offense of overconsumption. The intent is also to bring about a more responsible consumption behavior among that one percent, then the next one percent, etc., and in the process reforming consumption behavior among a very large segment of the American (and other) citizenry.

Elitists won't like this book. Maybe libertarians won't either. It matters little in either case. But Big Money is different. Big Money matters because it has the means to squash movements from the get-go. Big Money may try to squash this book, and the steady state revolution, by pointing a finger (in castigation, ironically) at those who would condone any kind of "class warfare," as they would surely call it. It's an easy thing for them to do, politically. It's an unfair rhetorical ploy—no one is promulgating violence or even vandalism—but since they are likely to use it anyway, let's look a little closer at what it means and how far they might take it.

In any situation where a growing class of people is coming to recognize and abhor a particular behavior of another class, and that class is accustomed to wealth and privilege, it will do what it can to stem the tide of change. It took tragically longer than it should have for slave-owning to be abolished, because many slaveholders were getting filthy rich from slave labor, and they fought tooth and nail to uphold the institution of slavery. In the US, for many decades the fight took place primarily in civilized social venues such as town hall meetings and newspaper editorials. When slaveholding became sufficiently revolting to a sufficient percentage of Americans, they began to win the war of words. Slaveholders, though, refused to align with the evolving social mores, and we all know the rest of the story. Slavery was ultimately abolished, but only at the cost of catastrophic levels of American—and African—blood and treasure.

But by now, when someone takes a slave in the US, that person is considered a criminal and is thrown in jail.

No doubt, Big Money would like to squelch this book by framing it as an irresponsible or even evil attempt to foment a violent, class-based revolution. But no; it's been clarified already and I will reiterate it yet again: violence is denounced in the steady state revolution. Not only is violence a bad thing to commit, it's not necessary for a successful steady state revolution. In fact, violence would impede the steady state revolution because, to be successful, steady staters must occupy the moral high ground, which is off-limits to violence. The steady state revolution requires a moral awakening to the technical knowledge that the ecological footprint of the liquidating class is so inequitable, and so damaging to the kids and grandkids, that it is a very bad thing. Just because the Civil War was the culmination of a particular moral awakening doesn't mean that all socially-based revolutions must turn violent.

I've already mentioned smoking. Class-based violence against smokers wasn't necessary to bring about a dramatic reduction in smoking. Rather, it's been a non-violent revolution, a castigation of the smoking class by the non-smoking class, that has given us the collective smoke-free public space.

Anti-smoking and anti-slavery: two class-based revolutions with major results, one entailing unprecedented levels of violence and one with virtually none. The steady stater is no more likely to shoot a liquidator than the non-smoker is to shoot a smoker. But just as the slave-driving tobacco peddlers of yesteryear gradually, and then suddenly, lost the public's respect and then acceptance, so too should the liquidators lose their place in society. With economic growth at the crossroads, such reckless and greedy behavior does not warrant the privilege of citizenship.

Laws should be passed to prevent citizens from exceeding certain levels of consumption, say one-tenth of a liquidator's footprint for starters. But now that we're moving from horse to cart—from paradigm shift to policy reform—it's time to consider a systematic approach to policy reform.

CHAPTER 11

A Call for Steady Statesmen: Policies for a Full-World Economy

> *Four in five Chinese believe protecting the environment should be a priority even if it means less economic growth.*
> PEW RESEARCH CENTER

> *China will try to slow GDP growth to ease pressure on the environment following a series of unusually stark warnings from senior ministers about the country's current mode of development.*
> THE GUARDIAN

WHEN A SUFFICIENT proportion of citizens and policy makers have come to recognize the everyday inconveniences as well as the extraordinary dangers of further economic growth, the time will have come for serious public policy reform toward the steady state economy. Hints of this awakening have appeared. For example, in 2011 China decided to moderate its economic growth rate from nine to eight percent. China's decision was newsworthy not so much for the intentional tempering of the growth rate, which many countries have done at times to prevent inflation. It's also true that eight percent is still a furious rate of growth. What is newsworthy, however, is that the Chinese government explicitly tied the lowering of their growth rate to environmental protection. In an online chat with Chinese citizens,

FIGURE 11.1. Dust and haze over the Yellow Sea and eastern region of China, inland to Beijing in the North, October 20, 2012. The Chinese leadership has acknowledged the conflict between economic growth and environmental protection. Credit: NASA Earth Observatory

Premier Wen Jiabao said, "We absolutely cannot again sacrifice the environment as the cost for high-speed growth."[1] This qualifies the decision as a precedent for steady statesmanship.

However, and in general, citizens and politicians worldwide do not yet identify the numerous threats of economic growth *in terms of* economic growth. Not even close. Instead, threats such as global warming, pollution and biodiversity loss are seen as technological shortcomings, diplomatic deficiencies or mistakes to be grown around. This is especially true in the US where, despite the gaudiest living standards ever enjoyed by a citizenry, and despite all the evidence for an overgrown economy, economic growth remains one of the highest priorities in the domestic policy arena. This has been true through thick as well as thin.

Although the time has not quite arrived for policy reform, the time is definitely ripe for scouting the policy options that will be increasingly sought in the context of Peak Oil, climate change, high unemployment and financial crisis. The simple act of talking about such options creates political space for policy tables to be set. Furthermore, if these policy options are not discussed now, the danger is that we will have all the wrong responses to Supply Shock. For example, as Peak Oil triggers stagflation, and policy makers seek answers, what should we expect them to do if the only game in town is still economic growth? Of course they will push even harder for developing other energy sources. Sure, this will also quicken the development of "green growth" sources such as solar and wind power, but as we saw in Part 3, this is really a strategy for less-brown growth, and we're at the point where we can afford very little more browning of the environment. Furthermore, to the extent that economic growth is the goal, and that less-brown sources will be insufficient to maintain that growth, the obvious outcome is the proliferation of dark-brown and fast-brown sources such as coal, tar sands and shale oil. Insidiously and profoundly dangerous nuclear power will be pitched as "green" in the context of global warming,[2] while Big Money convinces millions that those who warn of nuclear danger are just tree-hugging worrywarts. Indeed we are seeing all of these trends already, for Peak Oil is real and the economic margin is a ruthless force, pushing the economy into previously protected areas and into evermore dangerous options.

So, in terms of economic policy, step one in protecting the planet, ourselves and the grandkids from the juggernaut of economic growth is adopting the right goal. Fortunately, the basic alternatives are easy to identify. With economic growth at the crossroads, there are but two alternative paths: recession and the steady state economy.

The fact that there are only two alternatives to economic growth is worth dwelling on a bit. Invariably, when the pursuit of economic growth is criticized, some will immediately question the critic's belief in mom, apple pie and (if you're an American) Chevrolet. If you're not for economic growth, you must be a communist, or an anarchist at best. Or you're for "shutting down the economy." These kinds of responses must be anticipated and immediately revealed as reactionary in the extreme, lest the discussion be derailed in a heartbeat. When the "communist" charge is leveled, we need only point out that communists and their governments have pursued economic growth as ruthlessly as Wall Street, and with the same environmentally destructive results. It's not communism, socialism, capitalism or whatever-ism the steady stater seeks, but rather environmental protection, economic sustainability, national security and international stability. Nor is anyone, at least anyone sane, talking about "shutting down the economy." We are talking about the process of economic *growth*, not the existence of economic activity. To put it as simply as possible, when something is defined as an *increase*, whether it be in temperature, awareness or GDP, there are only two alternatives: a decrease or a steady state.

So clearly, the first step in policy development toward a steady state economy is adopting the steady state economy as a goal. Once we have the right goal, the other aspects of policy design fall into place.

Political scientists provide us with a general framework of public policy denoted as "S → A → T → G," where S is a policy statement (such as a statute or executive order), A is an agent (such as a government agency), T is a target (a group whose behavior will be influenced), and G is the goal.[3] For example, your town may have an ordinance (S) saying the police (A) will ticket you (T) if you spit on the sidewalk, in order to keep the sidewalk clean and sanitary (G). Although listed last, it is the goal that drives the formation of the whole policy framework. Without the goal, no S, A or T would exist.

But is economic growth really stated as a goal, in and of itself, or does it simply occur as a result of population growth, consumer behavior and numerous lesser economic policy goals? Sometimes it is a policy goal per se, and we will explore a few examples, but more importantly, if we take away the "per se," then it is clear that economic growth is one of the biggest goals ever to occupy the policy arena.

Going back to the sidewalk-spitting example, the ordinance doesn't state explicitly: "The goal is to keep the sidewalks clean and sanitary." It doesn't have to, because the goal of clean, sanitary public conditions are probably spelled out somewhere else in the town's code. Even if sanitary conditions are not mentioned anywhere in the town's code, such conditions are implied in policies calling explicitly for "public health." And even if there are no public health policies, frankly, it would be a matter of common sense. Certainly a very tiny minority, if any, wants to encounter spit on the sidewalk. The spitting ordinance was adopted as town council members thought with common sense about the various threats to clean and sanitary sidewalks. The pursuit of clean and sanitary sidewalks motivated the council to adopt the ordinance, which called for police to ticket spitters.

During the Reagan Administration, several federal agencies (big As), including the Army Corps of Engineers and the US Forest Service, had their missions redefined to include "economic development" per se. Although great care is taken in ecological economics to distinguish between economic development (a beneficial change in economic conditions) and economic growth (a quantitative increase in the size of the economy), such is not the case in political and bureaucratic circles. Indeed, the conflation of growth and development is the primary reason why ecological economists are so insistent on distinguishing between them to begin with. But conflated they are, and "economic development" in a mission statement is a license to encourage and contribute to economic growth.

Now when you are the commanding general of the Army Corps of Engineers or the chief of the US Forest Service, with all your deputies, assistants, other political appointees and sundry bureaucrats, virtually everything you do is geared toward achieving, facilitating or at least not obstructing economic development. That's the way it should be, given your mission, and you're not above the fray in conflating development with growth. In fact, it's likely you're not even aware of the distinction between growth and development. So when you approve a policy by which you will steer, let's say, a timber company, you better be able to explain how it contributes to the goal, which in this case you could call not only "G" but GDP.

See what the Reaganites got away with? Growthmen at the helm can do a lot of lasting damage in a short period of time. Once economic growth or economic development is embedded in a mission statement, it's not easy to expunge. Today, the Army Corps of Engineers' mission is to "provide vital public engineering services in peace and war to strengthen our Nation's security, energize the economy, and reduce risks from disasters." "Energize" is one of those verbs that, along with "stimulate" and "spur," is often used as a synonym for "grow."

Going back to our S → A → T → G model, in the US the biggest type of S is a statute passed by the Congress and signed by the President. Statutory law is the law of the land, trumping state and local laws and other policies. With the collective body of statutory law, the big, general A is, fittingly enough, the Administration. But of course most individual statutes identify one government agency, or a few agencies, to steer their targets toward a goal. For example, the Endangered Species Act tells the US Fish and Wildlife Service and the National Marine Fisheries Service to steer hunters, fishermen, loggers, miners and really a very long list of targets toward the goal of preventing the extinction of a really very long list of species. Good luck! (We'll get to that in a minute.)

The ESA also happens to be an example of a policy for which the goal *is* clearly explicated: "To provide a means whereby the ecosystems upon which endangered species and threatened species depend may be conserved" and a few closely related aims. The goal was explicit because widespread species endangerment was a relatively new thing on the American landscape. People weren't accustomed to the idea that species might be going extinct all over, or even to the idea that it mattered in a lot of cases. Common sense hadn't yet evolved to encompass the widespread nature and repercussions of species endangerment. Unquestionably, the ESA was a progressive, precedent-setting statute. The philosopher Holmes Rolston III called it "one of the most exciting measures ever to be passed by the US Congress, perhaps to be passed by any nation."[4] It was also one of the most nuanced, especially among environmental laws, reflecting state-of-the-art science and fine tuning after two earlier versions were passed in 1966 and 1969. The relevance of this to economic growth at the crossroads will appear momentarily.

When it comes to statutory law pertaining directly and explicitly to economic growth, the most relevant is the Employment Act of 1946, especially as amended with the Full Employment and Balanced Growth Act of 1978.[5] The original and amended versions are commonly referred to in American policy and media circles as the Full Employment Act. Among other things, the Full Employment Act calls for "full employment and production, increased real income" and "balanced growth." Although the phrase "economic growth" is used nowhere in the act, phrases such as "increased real income" and "balanced growth" are essentially synonymous with economic growth, albeit with slight additional nuance. Any remaining doubt is eliminated by numerous other phrases and clauses in the act that clearly call for an increase in the production and consumption of goods and services in the aggregate.

By using the term "balanced growth," Congress has called for economic growth under conditions of general equilibrium. This means an economy growing in concert—an efficiently allocating, circular flow of money with no major eddies of unemployment.

Seemingly, then, the verdict is in: economic growth is officially a goal of the US government. Well, it's still not that simple. The S → A → T → G model does not stop with those four components. Rather, the authors of the model (Anne Schneider and Helen Ingram) describe how "rules, tools, assumptions, and rationale" are interspersed among the S, A, T and G. The key in this case is the assumptions underwriting the Full Employment Act. One obvious assumption is population growth. With a growing population, full employment requires economic growth. Given the assumption of population growth, then, the goal in this case may be interpreted as full employment and economic growth.

In policy analysis, historical context is extremely important. The historical context of the Full Employment Act was the Great Depression, during which unemployment not only devastated American society, but shocked the pants off neoclassical economists. While they were busy pulling up their pants, Keynes strode through the mob, straight to the policy table. Of course, the British Keynes didn't literally stride to the policy table in the US; it would be more accurate to say that American economic advisors used Keynes's *General Theory* to *build* the economic policy table, at least the table where the Full Employment Act was drafted. Remember, prior to Keynes, the neoclassical economists didn't believe in a sustained or lengthy period of unemployment. They didn't believe in macroeconomic manipulation, and no one else knew any better, so there wasn't any macroeconomic policy table. Their pants kept falling down in the Great Depression, though, while Keynes's disciples were able to pull theirs up and move ahead for awhile. (Even the Keynesians' pants fell back down during the stagflation of the 1970s, but further reminders of hapless economists would be redundant given Part 2 of this book and other books such as *The Death of Economics*.[6])

The crucial point here is that population growth was a given, and given population growth, economic growth was required to achieve full employment. In other words, the real, primary goal of the Full Employment Act is not economic growth per se but full employment. "Balanced growth" might be a secondary goal, tacked on in 1978, but it is primarily a means toward achieving full employment in the context of population growth. Therefore, if population were stabilized, full employment would clearly still be a goal, while the pursuit of economic growth pursuant to the Full Employment Act would be an arguable endeavor. In fact, because too much economic growth results in collapse and high *unemployment*, the spirit of the Full Employment Act in the context of a full-world economy entails the cessation of population and economic growth. In other words, *in today's context, the Full Employment Act calls for a steady state economy!* Furthermore, it calls for a steady state economy at a level sufficiently within ecological capacity to ensure enough resources per capita to allow for full employment.

Unfortunately, the argument that the Full Employment Act calls for a steady state economy is not accepted—if it is even heard of—by neoclassical economists or politicians, for the various reasons revealed in Chapters 5 and 9. If this reasoning were widely accepted, the Full Employment Act as written would not be a barrier to the establishment of a steady state economy. Because the argument is not widely accepted, and will be fought by vested pro-growth interests, it will be necessary to amend the Full Employment Act to explicitly incorporate the rationale, in order to bring statutory law in line with a steady state economy. This would not be a complicated thing to do, technically. For example, the name of the act could be amended to "Full and Sustainable Employment Act." Within the act, "increased real income" would be amended to "stabilized real income." "Balanced growth" would be replaced with "sectoral balance" or "efficient allocation of land, labor and capital." Language would be added to state that the goal of sustainable, full employment requires stabilization of population and per capita production and consumption.

All of this would be quite straightforward and could be drafted by a smart graduate student with a nose for public policy. Such a student could draft the amendment in a political science course, for independent studies credit, or as part of a major paper or dissertation. For extra credit, or to impress the instructor for grading purposes, the student could also meet with the appropriate congressional representative and request that the amendment be proposed. Of course, if one student in the nation drafts such an amendment, especially if the student resides in the district of a dyed-in-the-wool growthman, the effort won't bear much fruit. But if numerous students in many congressional districts draft such amendments and meet with their legislators, you can bet the conversations in the hallways of Capitol Hill will buzz with this unprecedented expression of interest by young leaders toward a new vision that exudes common sense. The Full Employment Act won't be amended toward a steady state economy any time soon, but this new, palpable and exciting pulse of the electorate will have its effect "at the margins" as legislators adjust the fiscal policy levers, deal with the banks and negotiate international trade agreements. Their rhetoric and leadership will move away from growth at all costs toward sustainability, and perhaps toward other more healthy interests such as strengthening the family and getting along better with others in the world. Voters will be refreshed by this new focus on something other than "consumer confidence" and won't be as preoccupied with outspending the Jones. Indeed, consumer confidence will come to mean that consumers are confident in their ability to thrive without the newest gadgets, biggest cars and trendiest clothes.

Eventually, the Full Employment Act *can* be amended to call for stabilization of population, gross domestic product and jobs. Such an amendment would be of immense help in establishing a steady state economy. It is not the case, however, that the Full Employment Act absolutely *must* be amended for a steady state economy to be established. Laws of the people are important in the evolution of society but they cannot rescind the laws of thermodynamics. If economic growth remains an overriding societal and policy goal and population growth continues unabated, the Full Employment Act will eventually become patently impotent. Limits to growth will be encountered, making it impossible for balanced (or unbalanced) growth to continue. If the population is still growing at that point, runaway unemployment will ravage the grandkids as resources per worker decline, making it impossible for employers to hire more workers. At the extreme, the population will grow so large that only subsistence levels of resources will be available. At that point, with no buffer left for adjustment, a steady state economy will not be in the offing, but rather a collapse of Malthusian proportions. Some time after the collapse a steady state economy at a sustainable level may be pursued, but only if society has learned its ecological economics and retained enough governing capacity to avoid or recover from the chaos of collapse.

If society has not learned its ecological economics, then it is as doomed to repeat the pattern as lemmings in the Arctic. If it has learned its ecological economics but lost its governing capacity, dark ages of anarchy or feudalism may persist, and whatever governance does return may not be democratic in form. Theoretically, a powerful dictator could establish a steady state economy, but a dictator that powerful is unlikely to be benign.

It is of little use speculating further on post-collapse scenarios. The point is to try to establish a steady state economy prior to collapse, and here we are considering the role of statutory law in making this happen. We have demonstrated that economic growth will be limited even if the Full Employment Act is not amended toward a steady state economy. On the other side of the coin, the best possible steady state amendments to the Full Employment Act cannot *ensure* a steady state economy. Laws are hard enough to enforce when they are designed to prevent simple acts of incivility like spitting on the sidewalk. The bigger and broader the issue, the more difficult the enforcement becomes, and few in Congress actually expect a sweeping statutory goal to be met to a T (so to speak).

Now this may come as a revelation to many, but not if you've studied this book, or even the earlier part of this chapter: if one particular statute were strictly enforced, we would already have a steady state economy in the United States. That statute is the Endangered Species Act.

The ESA is perhaps the best example for demonstrating how important ecological economics is for a sustainable interpretation of statutory law. We saw in Chapter 9 that Congress was fully aware that "various species of fish, wildlife, and plants in the United States have been rendered extinct as a consequence of economic growth…" That doesn't mean Congress was aware that the conflict between economic growth and wildlife conservation cannot be reconciled through technological progress. Congress remained noncommittal on the prospect of reconciling economic growth with wildlife conservation by adding to "economic growth" the phrase "untempered by adequate concern and conservation." Nevertheless, a careful interpretation of the ESA, along with basic ecological principles, makes it clear that the ESA, were it fully funded and enforced, would indeed result in a steady state economy.

I have the benefit (and paid the costs) of having analyzed the ESA, word by word, along with much of its legislative history, for my PhD dissertation.[7] That analysis, along with a background in ecological economics, made it crystal clear that the ESA is a prescription for a steady state economy.[8] The prescription may be implicit, it may even be unintentional, but the ESA is a prescription for a steady state economy.

Section 4 of the ESA requires that species be listed as "endangered" if they are in danger of extinction or "threatened" if they are likely to become endangered. Once they are listed, they are to be protected, with the goal of recovery and delisting. Section 7 protects them from government actions (e.g., Army Corps or Forest Service "economic development" projects), and Section 5 protects them on private property. As far as the letter of the law goes, the ESA is truly some powerful stuff. Steven Yaffee, a renowned scholar of endangered species policy, called it "one of the most sweeping pieces of prohibitive policy to be enacted."[9] Bill Reffalt, a long-time leader with the US Fish and Wildlife Service before and after the passage of the ESA, called it "the most far-reaching wildlife statute ever adopted by any nation."[10]

The problem for growthmen is this: when the causes of species endangerment in the US are scrutinized, it eventually becomes apparent that behind these causes are a veritable *Who's Who* of the American economy. The causes of endangerment can be broken down into finer categories—I used 18 in my dissertation—but roughly speaking they include agricultural, extractive, manufacturing and service sector activities, plus the development and maintenance of economic infrastructure (roads, power lines, canals, etc.), economic byproduct (pollution), and various incidental effects of economic growth, such as climate change in a 90 percent fossil-fueled economy and the introduction of invasive species in a world of international trade and interstate commerce. As noted in Chapter 1, "It *is* the economy, stupid!"

This linkage of species endangerment with economic growth is an extremely thorny problem for policy makers because a very high proportion of citizens believe economic growth is a good thing. The ESA may be "one of the most sweeping pieces of prohibitive policy to be enacted," but that's not necessarily saying much, when one of the most sweeping policy goals ever embraced, of any type, is economic growth.

But let's assume for a moment that the ESA could be enforced to the letter. What that could mean for the American economy was showcased from the get-go when the snail darter was listed in 1973, the same year the ESA was passed. The listing of this tiny fish required the powerful Tennessee Valley Authority (TVA) to halt construction of the Tellico Dam on the Little Tennessee River, because the US Fish and Wildlife Service concluded that the dam and its reservoir would harm the snail darter. TVA dams and reservoirs had long been the backbone of economic growth in the Appalachian region of the southeastern US, providing electricity to millions and creating conditions for urban and recreational development. The congressional delegation from Appalachia didn't take the listing sitting down. They proceeded to turn the snail darter into a poster fish for purposes of weakening the ESA. As Senator Howard Baker (Republican Senator from Tennessee) said on the floor of the Senate:

> Mr. President, the awful beast is back. The Tennessee snail darter, the bane of my existence, the nemesis of my golden years, the bold perverter of the Endangered Species Act is back.
>
> He is still insisting that the Tellico Dam on the Little Tennessee River—a dam that is now 99 percent complete—be destroyed…
>
> Let me stress again, Mr. President, that this is fine with me. I have nothing personal against the snail darter. He seems to be quite a nice little fish, as fish go…
>
> Now seriously, Mr. President, the snail darter has become an unfortunate symbol of environmental extremism, and this kind of extremism, if rewarded and allowed to

persist, will spell doom to the environmental protection movement in this country more surely and more quickly than anything else.

I am seriously concerned that if present trends continue, the Endangered Species Act will be perverted from its original intent as the means of protection of endangered species and be used instead as a convenient device to challenge any and all Federal projects.

If the snail darter can be found in the Little Tennessee River, there is a snail darter or some equally obscure creature in every river and under every rock in America. Opponents of public works projects will have a virtually limitless arsenal of weapons with which to do battle.

We who voted for the Endangered Species Act with the honest intention of protecting such glories of nature as the wolf, the eagle, and other treasures have found that extremists with wholly different motives are using this noble act for meanly obstructive ends.

That is precisely what has happened in the case of the Snail Darter against Tellico Dam, and if this perversion of the law is allowed to continue, the law itself will soon stand in jeopardy—and that will be the ultimate environmental tragedy.

We must not let that happen, Mr. President. The House has given us another opportunity to set things right, and at long last we should take it. I implore my colleagues to seize this opportunity to redeem our commitment to energy production while not forsaking our commitment to environmental protection, to turn away from extremism toward reason, to save both the darter and the dam.[11]

And thus was written another chapter, albeit a cute one, in the win-win rhetoric that we can have our cake and eat it too. We can save the snail darter while damming more rivers in the Southeast. While we're at it, we can save the salmon while damming more rivers in the Northwest. We can save the spotted owls in the Northwest too, while logging more Northwest forests, and red-cockaded woodpeckers in the Southeast while logging more Southeast forests. We can save the polar bears, tufted puffins, green turtles and picas while burning more fossil fuels. We can save all species while perpetually growing the economy.

The fact is that we can do none of these things. We saw in Chapter 8 how, due to the tremendous breadth of the human niche, the human economy grows at the competitive exclusion of nonhuman species in the aggregate.

But why spend so much time on environmental policy, indeed on one statute, when the subject is macroeconomic policy? There are two good reasons. First, it points out the crucial nature of getting the goal right. Second, it shows how a steady state economy can be brought about even without highly successful economic policies. These two reasons are closely related. If a steady state economy becomes an explicit policy goal with widespread public acceptance, then arguments such as Howard Baker's will lose effectiveness. Those who say, "We can't enforce the ESA any further because it slows down economic growth" will be overruled with the response, "Yes, of course the ESA will slow economic growth, and since the steady state economy has also become a policy goal, enforcing the ESA will help us achieve that as well as species conservation."

No one should envision full enforcement of the ESA as resulting in *Animal Planet*, though. Rather, ESA enforcement would apply the economic brakes at the margin. Where the American economy is on the verge of extinguishing another species, the relevant economic activities are not allowed to expand any further. A dam project here, a timber sale there, building permits, highway projects, oilfield development…across the country such projects would be foregone, little by little, until the human economy has settled into a certain balance with the economy of nature. It's an equilibrium in which we have a very full human economy coexisting with a long list of threatened and endangered species for whom we have drawn a margin in the sand. Such species will remain precariously perched on the evolutionary tree of life, and many will fall to the ground of extinction, but at least we won't be chopping the whole tree down at a rate of three percent GDP growth per year. With a balance of nature established, we can count on God, Mother Nature and evolutionary ecology to keep the tree alive, with new species gradually replacing the extinct while *Homo sapiens* finally expresses its sapience in the form of restraint.

The broader point is that environmental policy *is* economic policy, and that's the way it should be in a full world scenario. It is the natural policy outcome from a realization of ecological economics. If we are serious about economic wellbeing, national security and international stability, we better get serious about enforcing our environmental laws. That goes for clean air, clean water, biodiversity conservation, environmental impact assessment, sustainable forest management, clean-up of toxic waste—all policies that contribute to maintaining and restoring ecological integrity and environmental health.

And we better couple that with reforming macroeconomic policy per se, or the pro-growth forces empowered by pro-growth policies will trump the effectiveness of environmental laws. The key is an explicit identification of the steady state economy as a policy goal, whether that be in an amendment to the Full Employment Act or with a superseding statute such as a Steady State Economy Transition Act.

Now let's take a look at some of the components—in addition to the goal itself—of a steady state policy program. An excellent framework is provided in the recent book *Enough Is Enough*, adapted from the proceedings of the first Steady State Economy Conference, held at Leeds University in 2010. *Enough Is Enough* identifies ten categories of steady state proposals, six of which are especially relevant to public policy, domestic and foreign. These include limiting resource use and waste production, stabilizing population, ensuring an equitable distribution of income and wealth, reforming the monetary system, securing employment and changing the way we measure progress. Three other categories have policy implications too, but are addressed more directly by citizens and NGOs. These include changing consumer behavior, rethinking business and production and engaging politicians and the media, topics covered at length heretofore. A tenth category, addressing global relationships, is most applicable in international diplomacy, which I will address alongside the issue of equitable distribution.

The broad category of limiting resource use and waste production overlaps substantially with environmental laws, but also includes policies that go directly to the heart of sustainability. The general idea of limiting resource use and waste production is self-explanatory but it will help to consider a few examples in some detail. The best example is a cap on fossil fuel extraction. The word "cap" itself connotes a steady state, and a fossil fuel cap is the best example because the global economy is approximately 90 percent fossil-fueled. Therefore, capping the extraction of fossil fuels would go a very long way toward capping the size of the economy and the ecological footprint.

The simplest approach is to cap barrels of oil, tons of coal and cubic meters of natural gas, starting at current levels of extraction. If necessary—and it probably is—these caps may be gradually lowered for purposes of fitting the economy to the planet. In other words, a certain phase of *degrowth* may be required prior to achieving a steady state economy that is optimal or even sustainable in the long run. This point warrants a short digression from our technical focus on resource-capping policies. (In a chapter on steady statesmanship, expect a mix of politics and policy.)

A growing understanding of the need for belt-tightening explains the political movement for degrowth in Europe. That's right, La Décroissance has become the rallying cry for a growing group of scholars, students and Green Party politicians. As a political movement, La Décroissance is closely linked to steady statesmanship because there is widespread agreement that the ultimate, long-term goal is a steady state economy, and that a certain amount of degrowth is necessary first. To be more precise, and for purposes of international equity and political stability, degrowth is called for in the wealthiest nations, coupled with economic growth in the poorest, but with a net effect of degrowth toward a sustainable global economy.

Decisions on whether to label a movement "steady statesmanship," the "steady state revolution," "La Décroissance," or something else are more decisions of political strategy than policy goals. We're all seeking the right-sized economy with social justice and efficient allocation of resources. In pursuit of these goals, surely the most politically effective choice of words depends on which part of the world you're in. However, and all else equal, labels that include the phrase "steady state" (in whatever language) are advantageous because such labels clearly identify the central, long-term policy goal. Also, when it comes to paradigm shifts, perhaps we should take them one at a time with our fellow citizens, who may not be willing to take them two at a time. When we take the step from economic growth to the steady state economy as a policy goal, it's only one more stepping stone to degrowth, and we have the momentum to get there quickly if need be. In contrast, the jump from growth to degrowth may be too daunting for the typical citizen to stomach, and in many countries it's not necessary.

Returning to the policy tool of capping fossil fuel extraction, we can also cap the amount of energy *used* to extract the fossil fuels or cap the acreage *used* for extraction. Caps are then enforced by issuing annual permits to producers who are fined if they extract unpermitted quantities, use unpermitted amounts of energy in the extractive process or use unpermitted acreage for extraction purposes. The initial allocation of permits should reflect the initial capacities and production levels of the extracting corporations or nations. This approach prevents unnecessary shocks to the market and is politically viable. Capping the fossil-fuel industry can also be kept as consistent as possible with a free market system by allowing corporations and nations to trade their permits or purchase them from one another after the initial allocation has been issued. For example, Exxon could sell some of its permits for the extraction of oil in the US to Shell. This is an example of a cap-and-trade regime operating within a nation, pursuant to the laws of the nation. Broadening our geopolitical vision and pursuant to an international cap-and-trade agreement, BP could trade oil extraction permits to Gazprom, receiving in turn permits for the extraction of natural gas. Each firm would invest in resources based on market principles of supply and demand.

Of course such an international cap-and-trade agreement will not be forthcoming until steady statesmanship is well-developed in international diplomacy. Ideally, such diplomacy would be led by the wealthier countries who can most afford to undertake the transition to steady states at this point in history. To expand a bit on the horse-and-cart metaphor, the wealthy countries would be leading the international horses with carrots, or at least with a whistle of encouragement. In reality, such diplomacy will also require sticks; that is, impoverished nations calling out the wealthy to curb their unsustainable appetites while allowing for some much-needed growth among the ranks of the impoverished. Indeed, this

FIGURE 11.2. Precedents of steady statesmanship have been well-received in international affairs. The popular King of Thailand, Bhumibol Adulyadej (*left*) calls for the Sufficiency Economy, while Jigmi Y. Thinley, Prime Minister of Bhutan (*right*) advances Gross National Happiness, eschewing the conventional goal of GDP growth. Credits: (*left*) Government of Thailand; (*right*) Royal Government of Bhutan

trend has already commenced and is certain to intensify with global economic growth at the crossroads. This trend could eventually fit the model of the steady state revolution outlined in Chapter 10, but in this case the "castigation of the liquidating class" is carried out not by individual citizens within a country, but rather by nation states in international venues such as the United Nations.

There is something of a precedent already for a steady state revolution in international diplomacy. For example, the G77 is a coalition of non-wealthy nations that now includes 131 member states, and the G20 is a group of 20 self-described "developing" nations. These international blocs strive to improve their terms of trade with the wealthy, "developed" nations. They've had some success, too, but they haven't drawn any attention to limits to growth or the need for steady state economics in international affairs. They've basically had the attitude that "a rising tide lifts all boats, but ours should be lifted faster." No doubt they would make a bigger splash if they demanded a cessation of economic growth in the G8 and other wealthy nations in order to provide some growth capacity for nations in dire need of it. They would find support in La Décroissance and other steady-statish movements in wealthy countries.

Another problem with the G77 and G20 is that neither bloc represents exclusively steady staters. For example, both include China, which despite its recent tempering of GDP goals is this generation's symbol of national economic growth. The G77 also includes Middle East petroleum states such as the United Arab Emirates. Some of the worst examples in the world of liquidating behavior come from these Arab states, and for sustainability purposes, the ugliest example of all is the Mall of the Emirates in Dubai. The Emirates set out to become the quintessence of conspicuous consumption, and succeeded beyond their most unsustainable dreams. The Mall of the Emirates is epitomized by the Dubai ski resort, where wealthy Saudis, Swiss, Americans, Israelis, the Sultan of Brunei (who makes Schwarzman look like a tightwad) and whoever else has the money without the dignity can play in the snow while temperatures outside exceed 100° Fahrenheit.

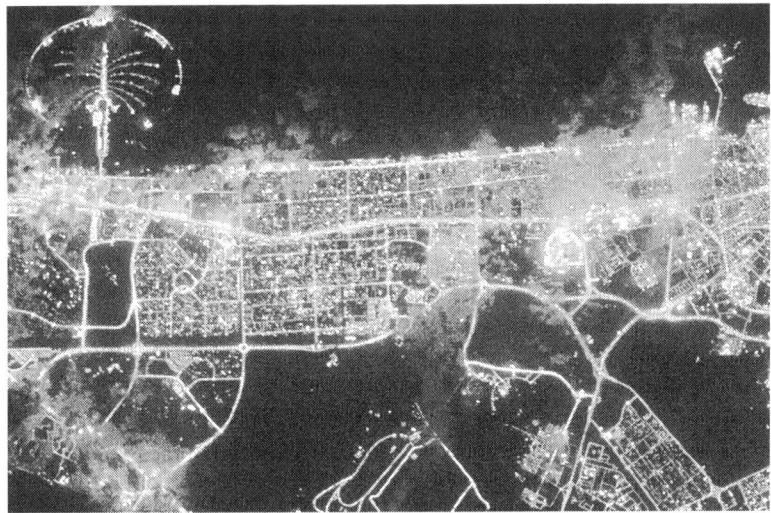

FIGURE 11.3. Dubai, geographic icon of the liquidating class, at night.
Credit: NASA Earth Observatory

For a legitimate steady state revolution in international diplomacy, a bloc of nations with the least-damaging GDP per capita is called for. There are 195 nation states,[12] so if we started with half the nations of the world and added a few to tidy things up, we'd come up with a G100 comprising primarily African, South Asian, Latin American, Eastern European and South Pacific island countries.[13] These nations would be united in diplomatically castigating the liquidating class of nations, which we might designate the G10: Qatar, Liechtenstein, Luxembourg, Bermuda, Norway, Singapore, Jersey, Kuwait, Brunei (thanks largely to the Sultan), United States and Hong Kong.[14] The G100 could carry out the precepts of the steady state revolution in ways not possible among individuals within a nation. For example, they could designate an annual Liquidator Nation of the Year among the G10, highlighting behaviors of its citizens like skiing in Dubai, driving Escalades or building mansions. Another approach would be to publish—and circulate widely—the ecological footprints of the liquidating nations in a matter-of-fact quarterly report. Yet another approach would be a boycott on trade with the liquidating class. Far beyond tinkering with the *terms* of trade, a G100 boycott would be announced as intending to lower the growth rates of liquidating nations (which it would) for the sake of global economic sustainability.

One of the beauties of a G100 would be its non-regional, non-ethnic, non-ideological character. The G100 would rise above historic, irrelevant conflicts such as North-South, East-West and capitalist-communist. The key, uniting variable would be sustainability of consumption. The most sustainable nations would be in, the least sustainable would be out. Sustainable nations would take pride in being so; unsustainable nations would be chastised as bad global citizens. Such diplomacy could only lead to a more sustainable global economy than the current one, in which nations race one another toward higher GDP. Certainly such diplomacy would empower the efforts toward cap-and-trade agreements, which alone would go a long way toward establishing a global steady state economy.

Cap-and-trade systems should start with fossil fuels but may also be enacted for all natural resources: timber, fisheries, minerals, etc. In fact, numerous marine fisheries are already managed pursuant to a cap-and-trade system in which the cap is called the total allowable commercial catch and the trading is of individual transferable quotas. The same key principle—limited extraction—applies whether the natural resource is renewable or non-renewable. The trade part is important for tapping into the allocative efficiency of the market and for making cap-and-trade a more politically viable solution. Powerful corporations populated by pro-growth free-marketers won't capitulate easily to capping, but the prospects for trading, at least, will appease them to some degree. The rest of the political lifting will have to be performed by policy makers who faithfully serve a public that understands the urgent need for steady statesmanship in an age of supply shock.

It will also help, especially wherever capitalism is favored, that capping is not needed throughout the economy. This follows from the trophic theory of money (Chapter 7). As long as we cap the producers at the base of the economy, manufacturing and service sectors will likewise be limited in scale. Capping the extraction of natural resources will also allow the fans and champions of the information economy to show their stuff, to prove to us that we can have perpetual growth without using more natural resources. Don't expect to see their stuff grow very much, though. The information economists will finally come to grips with the laws of thermodynamics, and that will be a good thing for all of us.

Capping and trading is no panacea, though. Not only will capping require a strong horse (a widespread paradigm shift away from economic growth), but the trading part will entail a lot of bureaucracy. The trading part is somewhat of a carrot to corporations, but not a particularly sweet carrot. It's trading, but not "free" trading. It must be overseen by a central authority, an "A" in our S-A-T-G model, and powerful corporate targets require equally powerful governmental authorities. This kind of trading is not "free" in the fiscal sense, either. One thing you have to grant to the free marketer is that, while the free market does a poor job of allocating natural resources fairly, it does so "for free." We do pay the unfair social costs—"environmental externalities" as they say—but not so obviously or directly out of our wallets, as in paying additional taxes explicitly to enforce a trading system.

The upshot for steady statesmanship is that we should strive for two things: to institute the necessary cap-and-trade policies *and* to avoid the unnecessary ones. The most necessary caps of all are for fossil fuels, because fossil fuels have a greater effect on growth rates than any other factor of production. Fossil fuels are the limiting factor for global economic growth at this point in history. By capping fossil fuel extraction, we make it less necessary to cap anything else. However, natural resources including fisheries, timber, certain minerals and groundwater in some regions, should also be capped. One reason is that there is no guarantee that fossil fuel caps will persist politically and therefore be enforced consistently. Also, in some regions, natural resources may be liquidated even in the absence of fossil fuel availability. Ironically, this will especially be so to the extent that we are successful in developing "green" (less-brown) energy sectors. So there should be a matrix of capping within the foundation of the economy; this will preclude the necessity of very much capping in the manufacturing and service sectors.

Meanwhile the avoidance of unnecessary capping is crucial for lowering costs, which we know (pursuant to the trophic theory of money) must be kept low enough, along with all other costs, as to be sustainable or payable. In other words, we cannot solve the sustainability problem by throwing evermore money at it—including into capping and trading administration—because increasing amounts of real money requires increasing the extraction of the very resources we are trying to cap!

This brings up the point that was first alluded to in Chapter 9—conservatives do tend to have one thing very right vis-à-vis sustainability. Deficit spending and mounting debt is unsustainable. Liberals tend to defend deficit spending, especially, on the grounds that it's good for economic growth, which in turn is supposed to be good for anything you can imagine. So we can easily envision such liberals supporting cap-and-trading in various sectors, yet inconsistently supporting deficit spending in order to enforce it all, boasting about the jobs to be created by spending a deficit, and most inconsistently of all propounding that the whole unwieldy mess will contribute to economic growth, thereby demonstrating that "there is no conflict between growing the economy and protecting the environment." Meanwhile the conservatives will be correctly railing against the mounting debt, but to what end? So far, it's all about "getting the economy back on track" and setting the stage for a renewal of economic growth. Clearly there are good intentions in both these camps; clearly these intentions are hamstrung by perpetual-growth economics (and pro-growth Big Money); and clearly the steady statesman must wed the good intentions from both camps with the implications of ecological economics to orchestrate a steady state outcome, including the judicious use of cap-and-trade systems.

While it won't be necessary to cap the production of most (if any) manufactured goods or services, it is important to cap pollutants at the other end of the pipe of economic production. By capping natural resources, we limit the throughput from the inflow end of the economy's pipe. Limiting throughput is essentially synonymous with limiting the ecological footprint and establishing the steady state economy. But due to the political and administrative difficulties of establishing and enforcing natural resource cap-and-trade systems, we should also cap the outflow of certain pollutants. Indeed, cap-and-trading regimes have their origins in pollution control, with the prototype being the sulfur dioxide cap-and-trade system originating in 1990. This system resulted from concerns about acid rain in particular, which made sense, but as with natural resource extraction, for the general purpose of steady statesmanship the idea is to cap emissions of pollutants that stem from fossil fuels. This essentially reinforces the capping of fossil fuels—the limiting factor for global economic growth—and has the extremely beneficial bonus of capping the carbon dioxide and other greenhouse gases emitted during fossil fuel combustion.

In many cases it will be more efficient to tax emissions than administer cumbersome cap-and-trade systems, especially for widespread pollutants that emanate from myriads of manufacturing sectors. However, steady statesmanship ultimately entails a stabilized tax stream, too, so that any increase in pollution taxes would be offset by income or property tax reductions. As Herman Daly says, "Tax bads, not goods." Speaking of Daly, a more detailed description of cap-and-trade systems and ecological or "green" taxes is provided in *Ecological Economics: Principles and Applications*, the excellent textbook by Daly and Joshua Farley.

One last thing about capping and trading, which applies to steady statesmanship in general: the basic solutions are not so complicated. The technical issues are challenging (see Chapters 5 and 8), the political hurdles are high and numerous (Chapter 9) and the widespread public paradigm shift is a prerequisite (Chapter 10), but crafting policy solutions requires little more than rolling up our sleeves and using common sense to "git 'er done," as they say. For example, it's easy enough to envision a Natural Resources Cap and Trade Act that would lay out the framework for which resources would be capped, how the caps would be set, how the permits would be allocated and traded and who would implement these regimes. This is yet another exercise that a grad student worth her salt could perform, contributing not only to her advanced degree but to the history of steady statesmanship. Likewise, it is easy to envision a Convention on Natural Resource Capping and Trading designed to address the global economy, hammered out with steady state diplomacy. The caps would first be applied in wealthier nations and the terms of trade would be designed to allow some convergence of impoverished nations toward standards of living enjoyed by the wealthy. The mostly-failed but well intentioned Kyoto Protocol would be worth revisiting as a starting point. Indeed, the Kyoto Protocol could yet be a successful tool in the policy cart, given the horse of a widespread steady-state paradigm shift.

Speaking of rolling up the sleeves and using common sense, the next policy issue for steady statesmanship is population stabilization. Nothing makes more common sense, with economic growth at the crossroads, than striving for a stable population. However, no other issue so exemplifies the horse-and-cart metaphor. Population stabilization stands no chance whatsoever of being addressed in national policies or international diplomacy as long as the overriding goal is economic growth. Recall from Chapter 5 that population growth is known in neoclassical circles as the key for perpetually increasing not only GDP, but GDP per capita, as more people must be devoted not only to consumption (for purposes of increasing GDP) but also to research and development (for increasing GDP per capita). But with the steady state economy as the goal of wealthy nations, and steady statesmanship a common theme in international diplomacy, population growth would be formally recognized as antithetical to economic sustainability, national security and international stability. Then it's time for rolling up our sleeves in the population policy arena.

There are three basic approaches to population stabilization: direct regulation, economic incentives and public encouragement. Direct regulation, such as China's one-child policy, is neither politically viable nor ethically acceptable in most cultures. Within a nation, it is coercive at best; internationally, it takes the form of war at worst. Perhaps the only place where direct regulation could play a legitimate and widespread role is neither within nor totally outside of a nation, but rather literally at the borders, where immigration policy is enforced. While open borders are conducive to freedom of choice, and constitute a generous policy of host countries, it must also be seen at this point in history that open borders allow for evermore overcrowding, or evermore overfilling of national economies. As this process of overfilling occurs in one nation after the next, these open borders are also conducive to a more-than-full world economy.

Wealthy countries are—and should be—brought to account for excess per capita consumption; likewise, overpopulated countries should be brought to account for excess demand on global resources. Many overpopulated and impoverished countries reject that charge, because often their plight has been caused or exacerbated by the plundering of corporations and governments from wealthy countries. No doubt they have a point there. The closest thing to a compromise of accountability, then, would be for wealthy countries to shut down their borders *in proportion to* their slowing of GDP growth. In other words, a wealthier country announcing and undertaking the transition to the steady state economy would be justified in shutting down its borders, and supported in international diplomacy for doing so. On the other hand, for a wealthy country to shut down its borders while pursuing globalized trade and economic growth would appear greedy with poor immigrants at the door. That's because it *would* be greedy. Such a nation would be shunned by the international community, which in today's world is ultimately a threat to national security.

Meanwhile, once the wealthier countries have undertaken the transition to steady state economies, the onus will fall upon impoverished countries to stem the rising tide of misery by doing everything ethically possible to slow their population growth and lessen the emigration pressures on wealthier countries. Because

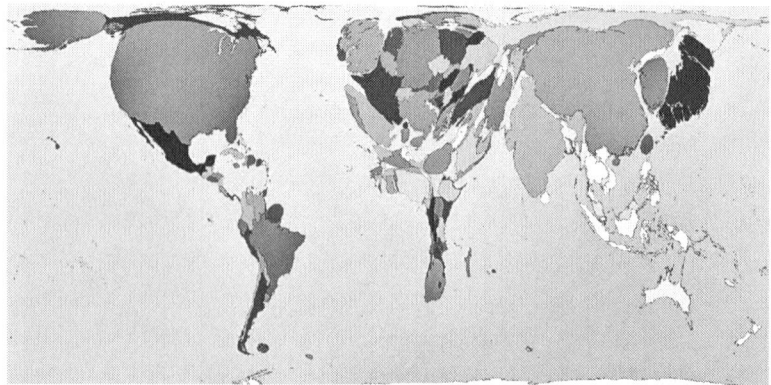

FIGURE 11.4. An ecological footprint map of nations (*top*) and the UN Economic and Social Council Chamber, a promising venue for steady statesmanship. Credits: (*top*) SASI Group (University of Sheffield) and Mark Newman (University of Michigan); (*bottom*) Mark Garten

population growth rates decline under conditions of higher GDP per capita (the "demographic transition"), it would behoove wealthier countries to assist impoverished countries in general and especially with population stabilizing efforts such as family planning education and the education of young women. Any amount of sacrifice by a wealthy country in order to assist with population stabilization in overpopulated countries will generate goodwill—good for goodness' sake and for generating the political capital needed for closing down borders.

Within a nation, the primary economic incentives for stabilizing population will be found in the tax code. The first step, then, toward stability in the US is to search the codes (state and federal) for existing growth incentives. The lowest-hanging fruit is the tax credit for child dependents. More than most, this is an example of a policy reform already in the cart, just waiting for the horse. The horse in this case is not only a polity supporting the steady state economy, but supporting it out of concern for the child dependents of tomorrow. Eliminating the perversely unsustainable tax credits for having more children today is a policy reform awaiting true steady-state leadership, and the first policy maker to push this reform into prominence will play a historical role in population stabilization.

Once such tax credits are eliminated, it is only another step to institute a progressive tax *debit* for child dependents, "progressive" meaning that the debit increases with each additional child. Essentially, we would be increasing the marginal costs of childbearing in order to decrease the demand on national and planetary resources. A demographically progressive tax code would not only engage economic incentives, it would also send a strong signal every year at tax time that the country found something to discourage in having too many children. In other words, it would contribute to or reinforce the country's paradigm shift toward a steady state economy.

I will be the first to admit a certain distaste for treating couples as a disembodied "T" in the S-A-T-G framework. Born and raised a Catholic, it runs against my grain to talk about any authority collecting money to reduce the propensity to procreate. It makes me uncomfortable and vaguely nauseous to think in such economic terms as increasing the marginal costs of childbearing to decrease the supply of children. Yet, also as a Catholic, I was raised to respect nature, to care about the health of the living in my home country and abroad and to be concerned for future kids and grandkids. None of those do I do well, if I don't make an ethical effort toward population stabilization today. And what makes me more nauseous than taxes on childbearing is the fullness of the world we explored in Chapter 1.

I think it is worth mentioning that I do not have any children. There are several reasons, but at least *one* is an awareness of the damage our ecological footprint is causing. I don't particularly enjoy sharing this personal information with readers, and I'm not holier than thou, but for those who think about the consequences of having children, it's important to know that you're not alone. For those who haven't thought about it, now is the time to start, especially if you're a politician. A growing number of us don't like voting for people with four or five kids because we think that level of resource commandeering by one family is greedy. Think of the "Octomom."

This brings us to the third approach to population stabilization, which is public encouragement. As with establishing a steady state economy, the key to effective encouragement is identifying the goal with clarity to begin with. When a nation decides to undertake the transition to a steady state economy, it should simultaneously adopt a formal policy of population stabilization, with a target date for the achievement of stability at some approximate level. In other words, this population policy should be part of the amended Full Employment Act or Steady State Economy Transition Act mentioned above. Even if there are no teeth in the legislation, establishing the policy goal is itself a form of public encouragement. It will instantly legitimize each subsequent policy reform toward population stability. If it should become the case in the course of political events that population stabilization becomes a viable goal prior to acceptance of the steady state economy—for example if the information economy rhetoric leads the country to believe that economic growth is sustainable even if population growth is not—then by all means the opportunity should be taken to formalize population stabilization as a goal of the polity. To the degree that this goal is attained, at least one of the two crucial steps (the other being stabilizing per capita consumption) in steady statesmanship will be accomplished. This is highly unlikely, though. Almost certainly, stabilizing population will only become a politically viable goal once the goal of a steady state economy is accepted, and not a minute before.

Beyond setting the right goal, public encouragement toward population stabilization means political leadership and public education programs that raise awareness about limits to growth, the need for a steady state economy and the essential role of population stability for a steady-state outcome. In the US, for example, the nation's population and its growth rate should be announced in the annual state of the union address. The President should express appropriate concern about the pressures on the environment and the capacity to sustain natural resources for future Americans. Any progress toward stabilization should be commended. This alone would go a long way toward feeding the horse of public opinion, and politicians at all levels would find it much easier to encourage citizens to have one or two instead of three or four or more children.

Meanwhile, public education programs should begin during primary education and should appear in community education programs designed for social welfare and basic home economics. The primary message should be about limits to population growth and the need to save room for future generations—"breathing room economics" as Rob Dietz calls it.[15] With that message at the core, specifics about family planning, financial incentives for small family size, caring for a single child and related subjects can be added.

If this doesn't sound particularly convincing as an approach to population stabilization, it should help to recognize that it doesn't have to be, at least not in most wealthy nations where the "native" growth rate of existing citizens is nearly already stable and where most of the population growth is coming now from immigration. Therefore, with a legitimizing goal of a steady state economy, only a little progress is necessary to bring national population growth rates down to stability. But again, this also assumes that the nation has set up the "no vacancy" signs at the borders. And to help in stabilizing global population, the wealthy nation must be prepared to participate in a full program of steady state diplomacy to address natural resource extraction, fossil fuel emissions, population growth and per capita consumption.

Which brings us to the issue of distribution; that is, the distribution of income and wealth. Recall from Chapter 6 that while the old maxim, "a rising tide lifts all boats," had some merit in an empty world economy we know this approach is defunct in a full world. The tide can only rise so far, and there is only so much material for boat-building. So we would like the wealthy owners of luxurious yachts to share a bit, especially if they haven't done much to earn those yachts. But we do not want them to be attacked by waves of poor pirates, nor do we want so much rage at sea that gunboats are sent in to settle the matter. Not only would there be innocent casualties caught in the crossfire, but none of that is sustainable; it uses a lot of boats and pollutes the seas.[16] What we want, in other words, is an equitable distribution of wealth, embraced as a feature of steady state economics. That's when we can expect a legitimate coast guard to police the relatively calm waters, arresting the occasional bona fide pirates, be they scallywags or CEOs or both.

One approach to fairness is the steady state revolution described in the previous chapter. Although the steady state revolution is mostly about lowering the liquidator's propensity to consume, it is also conducive to a more equitable distribution of wealth. Some of the income that would have been spent on profligate consumption is instead spent on public improvements (such as parks, arts and educational endowments) and direct, redistributional charities.

The steady state revolution has two things in common with a free market: it reflects consumer preferences and it comes without government intervention. That doesn't mean a steady statesman couldn't participate. Indeed, it would be hard to imagine a greater contribution than a president addressing overconsumption in a State of the Union Address. Can you almost hear it, almost see it? "This year our wealthiest citizens—the upper one percentile—reduced their consumption by seven percent. They're still doing fine, mind you. [The President grins.] Meanwhile, our consumers on the lowest rung were able to increase their purchases by eight percent and donations to public causes increased by six percent. These are trends we should appreciate and encourage. [The President leads a round of applause.] Along these lines, there are some policies that will dovetail with these trends and help us to achieve sustainability for our kids and grandkids…"

That's when the bold president could—or we citizens could even sooner—call for a cap on income or a cap on wealth. We have already explored the concept of capping natural resources and emissions. We also know that real money represents the ecological footprint (Chapter 7). Too much money can't fit on the planet. Likewise, too much money can't fit in a country, a county, or a city—certainly not equitably for people elsewhere. Too much money here means not enough money there. All we need to do is extend this logic to the corporate board or the household and we're talking about caps on salaries and wealth.

In the US we've seen a remarkably successful system of salary capping—the National Football League salary cap. The NFL salary cap has done more to keep football, American style, alive and well than any Peyton Manning pass or Devin Hester dance. Rather than the richest CEO buying the Lombardi Trophy year after year by assembling the highest-priced players, we have legitimate competition among 32 teams. Rather than a disgruntled fan base, disgusted by unscrupulous CEOs, we have a league of fans who (for the most part) respect each others' traditions and teams. We have historical antiquities (at least by the standards of American sports) such as Lambeau Field in Green Bay, Wisconsin, which without an NFL salary cap would have been replaced by a cheese factory or a Kmart. Meanwhile the Green Bay Packers themselves—loved by many for their small-town story—would have been sent packing, perhaps literally, back to the packing plants for which they were named.

Of course precious little else about the gaudy NFL is sustainable, but at least the salary cap proves to us that not only is such a thing possible, it can be wildly successful at leveling the playing field and keeping the fans interested, engaged and civil. We need to move from the NFL salary cap outward in American society and downward in level to most occupations. We need to get to where the ecological footprint of all that money fits within the nation's environmental capacity, while still keeping its citizens happy and healthy. We want the citizens of soccer-playing, rugby-playing, and cricket-playing countries to be healthy and happy with us, too.

Salary is only one form of income—probably the most actually earned form—and what we are really after is capping *gross* income, including rents, profits and interest. Given how tightly we track income in wealthy countries (think of the Internal Revenue Service in the US), administering such a system would be fairly straightforward. Capping can be administered by prohibiting payments beyond a certain threshold or by taxing income beyond the threshold. Tax revenues would then be used for public purposes and, where necessary, as a safety net to bolster the incomes of the poor. All the arguments about "welfare" and engendering a "welfare class" have already been made in other books, by think tanks and in political campaigns. No rehashing is necessary here, and clearly it is more important to cap incomes than to provide minimum incomes. Politicians must ensure that tax revenues go to public works that help the poor get by, even without direct welfare payments. Meanwhile, capping will invariably result in *real* trickle-down effects, not the tricky excuses used by supply-siders to lower taxes on the wealthy.

As for where to set the income caps, the key variables to consider are the ecological footprint of a real dollar and of course political viability. Earlier we noted the virtually criminal nature of impacting the planet with enormous ecological footprints. Theoretically, we can take the ecological capacity of the nation in terms of dollars, note its population and estimate a sustainable income at that population level. Then we have to decide how much variance from that income is socially appropriate and how politically viable it is to cap income at various levels. For example, there is plenty of research indicating that people are happier and cultures are stronger in societies that are more egalitarian. Some of the positive effects include better health, higher life expectancy, less problems with drugs and violence, less obesity and less incarceration. These findings provide powerful political leverage for capping incomes. Yet an absolutely equal distribution is neither desirable nor politically feasible. So the question is, how much larger than minimum incomes should maximum incomes be?

How about fifteen times as large?

See, it's not really so hard, is it? Certainly it is not hard to start with *something* on the policy table. Furthermore, it's likely that this fifteen times proposal resonates with a substantial share of readers. It somehow seems quite commonsensical, no? At least for occupations in similar sectors, right? If you're a barber working 40 hours a week, you might think it okay for another barber across town, also working 40 hours a week, to make a little more (or a little less) than you. Now maybe if you're in Pulaski, Tennessee, and the other barber is in New York City, you could understand if the other barber makes twice or even three times as much as you. Of course that barber may have other sources of income, too. He might have a mutual fund or he might rent out a room in his condo. So you can probably understand if he makes even ten times as much as you.

Now this barber, he may have also inherited money from his uncle the banker, who he always hated, or maybe he won a lottery one night on a drinking binge at the casino. Maybe when he sobered up he invested all that money on Wall Street (a somewhat safer casino, usually) and now he makes 50 times as much as you. Now you're starting to wonder how fair it is, and if you've undergone the steady-state paradigm shift you're wondering how sustainable it is, too. What's that barber doing with all that income? Is he acting like that Schwarzman fellow and pulling out the rug from your grandkid's future?

So you're thinking it's one thing if that barber makes three or five or even ten times as much as you, but 50 times as much is just plain wrong! He doesn't need anywhere near that amount, he didn't really earn it, and now he's turning into a liquidator. This is not working, not with economic growth at the crossroads, not with

Supply Shock upon us. You are a reasonable barber and one of a large majority of Americans (or Frenchmen, or Indians) who are honing in on 1,500 percent as a common sense, allowable order of difference from the lowest income to the highest, at least in a given sector. You can stomach another barber making 15 times as much as you, and you suppose somebody like an NFL player or a brain surgeon could make somewhat more than 1,500 percent of your income, but you're not too crazy about that either, and you definitely don't like the idea of bailed-out bankers and plastics CEOs making 1,000 times as much as you, or 100,000 percent of your income. To say that you're not alone is a major understatement, meaning there is plenty of political viability for caps on income and wealth.

There are many sectors, salaries and other sorts of income to consider in developing caps on income and wealth. In fact, we haven't dealt much with wealth per se and policies such as inheritance taxes. Clearly the subject matter warrants a whole book and probably numerous books. Developing detailed proposals is a job for think tanks, policy entrepreneurs, progressive politicians and, once again, grad students. We've seen enough here to serve as part of a steady state policy framework. One common theme has been the trophic theory of money, which tells us that real money supplies and flows must be stabilized to be sustainable. That almost brings us to the issue of monetary reform, but first a bit more on the role of grad students is in order.

Developing detailed proposals for steady state policies is not yet in the cards for most think tanks, policy entrepreneurs and progressive politicians. Trust me, it's tough to find funding for steady state think-tanking, much less advocacy. Big Money tends to be pro-growth, of course, so in steady-state circles there is nothing analogous to the Cato Institute, nor for that matter the Brookings Institute. This may never change to a substantial degree, even if the general public undertakes the steady-state paradigm shift, because Big Money doesn't work for the general public. But a significant share of the funding in the university system comes with no strings attached.

In academia, scholars and students left and right are recognizing the disconnection between conventional economics and the state of the planet. So far the response has been a proliferation of ecological microeconomics; that is, estimating the value of natural capital and ecosystem services. Several institutions have become especially known for such research, most notably the Gund Institute at the University of Vermont. But ecological microeconomics is only a marginal improvement over neoclassical economics. It's still limited to getting the prices right, albeit with a more complete accounting of costs. For all the reasons described in Chapter 6, getting the prices right is hardly an adequate response with economic growth at the crossroads. What we need now in academia is a flagship program for *ecological macroeconomics*, specializing in the trade-off between economic growth and environmental protection (a technical matter requiring ecology and economics) and steady statesmanship (a policy matter requiring political science and sociology). This flagship will be something of a complement to the Gund Institute (with its micro-focus) and a counter to the Chicago School (with its neoclassical pro-growthmanship). The potential for such a presence is palpable at several universities, including Michigan State University in the US, Leeds University in the UK, and the Autonomous University of Barcelona. At these universities and others, grad students can already seek degree programs geared toward steady statesmanship and can focus their masters theses or PhD dissertations likewise. Departments that are starting to sponsor such research include geography, political science, sociology, environmental science and, at some schools, even economics. The entire policy framework provided in this chapter can be fleshed out by such graduate research, and many of the graduates themselves can go on to be the steady statesmen and women we need in the political pulpits and at the helm of public policy.

Regarding the policy framework provided in this chapter, we next address the monetary sector. With the exception of population stabilization, nowhere is the principle of putting the horse before the cart more important than in monetary affairs. Monetary policy is currently focused on preventing inflation while stimulating economic growth. The primary tools for pursuing these goals are the money supply and the interest rate. Basically, increasing the money supply and lowering the interest rate are conducive to economic growth, but also conducive to inflation. So monetary authorities (such as the Federal Reserve System in the US) attempt to stimulate the economy without causing inflation, and it's a delicate dance.

It's also a tangled web they weave, these monetary authorities, as they inevitably get caught up in the broader and wackier world of finance, private as well as public. I do not recommend Andrew Ross Sorkin's *Too Big To Fail*, as it would take too many hours of your life (as it did mine) to plow through the 600-page minutiae of incestuous dealings among financial and monetary titans such as Alan Greenspan, Henry Paulson, Ben Bernanke and (alas) the world-class liquidator Stephen Schwarzman. (In fact, *Too Big to Fail* is loaded with liquidating lore, but the marginal utility of such information diminishes rapidly.) I recommend instead that you take my word for it, along with your common sense. Stocks, bonds, insurance, mortgages and increasingly surreal derivatives with "collar," "strangle" and "iron butterfly" options constitute a shape-shifting matrix that challenges the monetary authorities' abilities to stay plugged into reality. By "reality" I mean the real economic sector with its trophic structure of agricultural, extractive, manufacturing and (non-financial) services sectors. The volume of financial transactions on such products as "rainbow derivatives" in no way reflects the actual production and consumption of real goods and services. This is why stock markets, mortgage markets and the financial markets in general can boom and bust like balloons at the county fair while the economic capacity of Planet Earth stays approximately the same, punctuated by the occasional volcano or meteor (and now threatened by trends such as climate change and biodiversity loss).

Because of the dubious connection between the real sector and the circus sideshow of the financial sector, even conventional, neoclassical economists have long questioned the effectiveness of monetary policy to stimulate the economy. The ironies never cease, for these economists see no real limits to economic growth, buying whole hog into the information economy and perpetual technological progress. Yet even they cannot imagine that monetary hocus-pocus can stimulate an economy to grow, whether by information or schminformation. Clearly they have a point, ironically or not, because when an economy has reached its real, natural, ecological limits, it doesn't matter what you do with the money supply or the interest rate. You can set the interest rate below zero, paying borrowers to borrow money, but you can't milk a dry cow. Mother Nature is constantly verifying this, from the lowest trophic level up.

Nevertheless, in the Keynesian tradition, it is just as obvious that monetary policy does affect growth rates when an economy is *not* operating at full capacity. Monetary authorities can indeed stimulate growth by lowering interest rates or increasing money supplies. But monetary policy doesn't have to be pro-growth. Carefully tempering money supplies and keeping interest rates from going too low clearly can *slow* rates of growth. Therefore, it is absolutely crucial to get our monetary authorities on board with the need for a steady state economy. As they make decisions affecting the rate of growth, they must increasingly recognize that decisions conducive to growth are "uneconomic" and cause more problems than they solve. Rather than prioritizing growth without inflation, they can prioritize steady statesmanship without deflation. And they can start taking pride in providing leadership toward a sustainable future.

If you think the idea of monetary authorities and financial gurus becoming steady staters is entirely beyond the pale, a few examples should make you think twice. Henry "Hank" Paulson was the CEO of Goldman Sachs—a Wall Street icon—before President George W. Bush called upon him to be Secretary of the Treasury in 2006. Serving as Treasury Secretary until 2010, Paulson has been a jet-setting mover and shaker of private and public finance, nationally and internationally. No one has been more representative of the financial sector in the 21st century. Pursuant to stereotype, we'd surmise him to be a Schwarzmanesque liquidator. But we'd be wrong, very wrong. Paulson is actually a "hard-core environmentalist" who drove (and presumably still drives) a Toyota Prius.[17] He's been a member of The Nature Conservancy for decades, has donated over $100 million to nature conservation projects and plans to donate his entire fortune to environmental causes.[18] While the Bush Administration infamously denied a human role in global warming, Paulson was a rare dissenter.[19] His family is likewise inclined toward environmental protection. His wife used to lead birdwatching tours for The Nature Conservancy. His son was on the Board of Advisors for the Wildlife Conservation Society.

While Paulson "was something of a baffling outlier"[20] by Wall Street standards, he is no less baffling by sustainability (non-Wall Street) standards. Paulson grew up on a farm in Illinois—he's *got* to know where the milk comes from. And get this: "Before college [Paulson] wanted to become a forest or park ranger. Instead he opted for a business career, getting an MBA from Harvard."[21]

I can't help thinking of my mom's certifiably Catholic admonition: "There but for the grace of God go I." Indeed, sheltered from neoclassical economics, I basically did what Paulson wanted to do, becoming a ranger, firefighter, biologist, etc. Instilled with principles of ecology, and with decades spent in the field, I wound up advocating a steady state economy when my PhD research led me to see the fundamental conflict between economic growth and environmental protection. Meanwhile, Paulson ended up instilled with neoclassical economics, Harvard-style, and devoted decades to economic growth. He eventually contributed millions of dollars to conservation, but I wonder if he ever pondered how the millions were generated. I also wonder what I would ponder with an MBA from Harvard, Stanford or the University of Chicago. In other words, I'm not standing in judgement of Paulson. Far from it, for Mom was right about "the grace of God." Rather than picking on Paulson, we should seek him out, connect on environmental matters and engage him in steady state economics.

Paulson's environmentalism is so dramatically ironic that I hesitate to offer other examples for fear of being anti-climatic. Yet every year since 1981, the Federal Reserve Bank of Kansas City has held an annual symposium in Grand Teton National Park.[22] Ben Bernanke, Timothy Geithner (Paulson's successor as Secretary of the Treasury) and a long list of other Fed and Treasury officials (many of whom are in a revolving door with Wall Street) gather in Jackson Lake Lodge to discuss the state of the monetary sector. Surely there must have been, over the 30 years of this outing, some notions of irony among these growthmen as they roamed the Teton trails after hours. Have none of them glanced at a Teton glacier and lamented its melting? Or spied a grizzly in a meadow, evoking thoughts of endangered species? Or had their peace disturbed at sunset by the sound of a Jake brake on Highway 26? Have none of these thoughtful men connected such disturbing thoughts with economic growth, the *summum bonum* of their careers? Surely some of them have, for few men are immune to soul searching. Steady staters worldwide should seek an audience with the monetary authorities and financial gurus, especially those with known Paulsonesque propensities, and solicit their steady statesmanship. Some of these authorities and gurus could surprise us with their solicitude.

Not only do the monetary authorities control money supplies and interest rates; they also have substantial control over banking regulations, including fractional reserve requirements. When you and other bank customers deposit your money, you must know that the bank doesn't keep all that money in the vault, in case you all want it back the next day. Rather, the bank assumes that few of you will need money the next day, and keeps only a fraction of all your deposits in the vault. The rest is loaned out to borrowers, at interest. It's not really the banks' money to loan, but they loan it anyway, in a sense creating money by fiat. It's all legal, this authority to create fiat money, and it nets the bank an easy income called "seignorage." This income is in addition to the interest paid by debtors.

In an empty-world economy, this was fine, at least for sustainability purposes. Most of the debtors were out working in the real sector, starting with the farmers and extractors toiling in the sun, wind and rain to wrest more of the Earth's natural capital. Producers needed money for tractors, oil rigs and boats; manufacturers needed money for mills, refineries and canneries. These debtors would sell their goods in the market, then dutifully pay back the bank the principal and interest.

It was fine for sustainability purposes, but of course the bankers made out like bandits. They accumulated income, giving them purchasing power, political power, philanthropic power, propaganda power, in proportion to the toils of labor and the resources of the planet. It's no wonder the apical ancestor of bankers, Mayer Amschel Rothschild (1744–1812), said, "Permit me to issue and control the money of a nation, and I care not who makes its laws."[23] Abraham Lincoln said, "The money power of the country will endeavor to prolong its reign…until the wealth of the nation is aggregated in a few hands, and the Republic is destroyed."[24] Henry Ford said, "It is well enough that people of the nation do not understand our banking and monetary system, for if they did, I believe there would be a revolution before tomorrow morning."[25]

If you're wondering how much fiat money bankers are allowed to create for themselves, that's determined by the fractional reserve requirement, which is set in the US by the Fed's Board of Governors. The reserve requirement varies by the size of bank and type of account. Reserve requirements for demand deposits (such as in checking accounts) range from zero percent (for small banks) to ten percent (for larger banks). Similar fractional reserves are required in many countries.[26]

To reiterate, banks are required to keep only a small percent of your hard-earned deposits available upon your demand. This in the wake of Enron, Bear Sterns and the incredibly unfair banker bailouts of 2008. If you're thinking this is as sustainable as a snowball in the Sahara, your optimism is reflected by the size of snowball you have in mind. The tiny fractional reserve requirement is less sustainable by the day, melting as it were in the context of global warming and all the other signs of a full-world economy. It needs to be raised.

In fact, true steady statesmanship entails phasing out the fractional reserve system entirely and replacing it with fee-service banking. As long as banks are allowed to issue new money by fiat, they essentially put the planet in debt and require natural capital payments. In other words and all else equal, fractional reserve banking assures us of uneconomic growth.

Fee-service banking is just what it sounds like: banks charge a fee for holding your money. The banks can package loans based on the receipt of such fees. Also fair game for lending are time deposits that, by definition, are off limits to the depositor for certain periods of time. The idea with banking reform is not to eliminate the practice of lending, but rather to make the rate of lending, and interest payments, sustainable. Banking reform is part of the broader macroeconomic policy reform toward a steady state. None of this reform is intended to shut down the economy, but rather stabilize it and make it sustainable. Even in a steady state economy, sustainable amounts of infrastructure and other manufactured capital will depreciate (pursuant to the second law of thermodynamics), and lending that enables the replenishment of such capital stock is necessary. Lending may also be required when one business starts up while another completes its run, or as one sector (such as solar power) gradually eclipses another (such as fossil fuels). In the context of stable populations, caps on resource use and pollution, and other criteria of a steady state, interest rates would presumably tend to reflect the rates of capital depreciation and business start-ups.

Another widely touted monetary reform is the establishment of local currencies. These are certainly legal in most parts of the world, including the US, where nearly every state has one or more local currencies. Examples include Asheville Dollars (North Carolina), Atlanta Hours (Georgia), and the aptly named REAL Dollars (Lawrence, Kansas).[27] Local currencies, by definition, are used within local communities by all who decide to use them as an alternative or supplement to the national currency. They have the huge advantage, with regard to sustainability, of de-globalizing the real economy, instantly lowering the energy and resource requirements of shipping, because producers and consumers are all local. As a store of value, they provide diversity and therefore resiliency in the monetary sector; no one wants to have all their "beans" in one pot. Another huge benefit is the community trust-building that occurs as the firms and households comprising the circular flow of money are actually friends and neighbors, or otherwise become acquainted as a result of local transactions. Imagine how it might feel to personally know who grows your wheat, bakes your bread, crafts your furniture and…banks your money. For most people, this knowledge adds something intangible to the quality of life. People feel more connected, together, united in advancing the welfare of the "village."

Local currencies may never become a prominent feature of steady statesmanship, because it takes a determined effort to launch and maintain them. Local currencies aren't a steady-state panacea, either; they too can be conducive to economic growth (minus the globalized aspect) if that is the community's goal. But in communities that have already adopted a steady-state policy goal, or at least undergone the steady-state paradigm shift, local currencies can be used to avoid the fractional reserve banking system with its growth imperative. Therefore, they are an important tool in the steady statesman's policy cart. Nothing will substitute, however, for steady statesmanship in the national monetary authority and in the legislative, administrative and ministerial bodies that set or influence interest rates, money supplies and banking regulations. Populating the congresses, chambers, parliaments, state houses and even the supreme courts with steady staters must be achieved; otherwise, the "margin" of the national economy will push like a bulldozer into any and all local communities, regardless of how sustainable they attempt to be on their own.

Next on our list of issues to be addressed by steady statesmen and women is employment. While some issues are more obviously connected to the establishment of a steady state economy—such as capping resource extraction or stabilizing population—no issue is more important than the maintenance of full employment. For the vast majority of people today, employment is required not only to make a living but to maintain a fulfilling identity and social network. Without the prospect of full employment, or at least very low rates of unemployment, the steady state economy is a political non-starter.

Earlier in the chapter I proposed amending the Full Employment Act to the Full and Sustainable Employment Act. I failed to note the handy, partial acronym "Full SEA," which with a bit of nicknaming license becomes the Full Seas Act. What a fortunate acronym it is, because "Full Seas Act" would have the tremendous upside of tapping into the metaphor of the rising tide—in this case having risen as far as sustainably possible—with every single utterance of the phrase.

Recall that the focus of the Full Seas Act was on stabilizing population because a stable population is a prerequisite to any prospect of perpetual full employment, and to a steady state economy. Shortly afterward, we looked at some basic policy tools for stabilizing population. In other words, we have explored much of the necessary terrain for maintaining full employment. But there are two major nuances that will face the steady statesman, and we must face them now. One is technological progress, and the other is the transition to a steady state economy, a period during which population may still be growing.

We explored the origins of technological progress in Chapter 8. However, we did not explore all the implications. One implication of technological progress is an increase in labor productivity. If you're new to this issue, don't let the lingo fool you. An increase in labor productivity doesn't mean the workers work harder. They're already working quite hard! Rather, as workers become coupled with newer, more efficient equipment and processes, more output is produced per hour of labor. This process of increasing labor productivity, resulting from technological progress, has invariably resulted in the layoff of workers, because capitalists have found it more profitable to substitute machine for man. Why keep a hundred ditch diggers with shovels when you can buy a gas-powered ditch-witch and hire one operator? (To put it in technical economics terms, the marginal physical product of manufactured capital has grown faster than the marginal physical product of labor, due to technological progress.)

Yet nothing physically requires the capitalist, or the government, to substitute machine for man. If the goal of full employment is to be reconciled with the reality of a full-world economy, the production of goods and services has to become more labor-intensive. Maybe it sounds bad, but just as increasing labor productivity doesn't mean the worker works harder, labor-intensive doesn't mean the labor is more intense. It simply means that the ratio of labor to capital is kept somewhat higher than the capitalist might opt for in an empty-world economy with no concern about environmental protection, economic sustainability, national security or international stability.

Clearly, we would all like machines to do the jobs, while we sit back and drink Margaritas or milk. But maybe it's not so clear if we're the ones out of a job. We—or the vast majority of us—have to recall that all that glorious, labor-saving technology is not owned by us. Wealthy capitalists own it, and while they may have nothing against you or your employment, they do have something against lowering their profit margins. If it comes down to hiring you or, more profitably, purchasing a robot, don't be surprised if the robot gets the nod.

The upshot is that it may become necessary to require a certain labor intensity of production. Such a requirement would be unfathomable in the absence of a steady-state paradigm shift. Although labor-intensity requirements wouldn't require state *ownership* (a key feature of socialism), they would entail a degree of central *planning* (the other key feature of socialism, but also of capitalism). Planning would be needed to ascertain how much labor would be required, in contrast to capital investment, to maintain full employment during the transition to a steady state. Yet the closer an economy gets to its ecological capacity, and the more perilous economic growth becomes, the less of a sacrifice some central planning appears.

Requiring a given labor intensity may be especially important in cases where the population is still growing as the steady state economy is adopted as a policy goal. Indeed this scenario seems highly likely. Most nations (or other polities) establishing a steady state economy as a policy goal will do so because, and while, trends in population and per capita consumption are obviously unsustainable. While the transition is being made to a steady state economy, and with populations still growing, efforts to prevent widespread unemployment will be essential. Such efforts must coincide with efforts to stabilize population, and both efforts must be successful.

Of course, labor intensity requirements would not be confined to the private sector. The closest thing to a precedent for labor intensity requirements in the US is Franklin Delano Roosevelt's New Deal programs during the Great Depression. For example, the Civilian Conservation Corps (CCC) did not go out and purchase as many bulldozers and chainsaws as possible, but rather hired as many shovelers and axe-swingers as possible. The same work was done (earth moving and timber cutting), but in a labor-intensive manner that reduced the problem of unemployment. Note especially that, all else equal, this approach to getting the job done is much less harmful to the environment because machinery and their fuels are not required, and people with hand tools tend to leave less severe scars on the land. Meanwhile the workers, especially young workers, find a certain healthy exuberance in outdoor physical labor, as long as it isn't overdone.[28] Similar labor-intensive programs have been administered in China, Russia and many other countries, but the New Deal is instructive in melding state-sponsored employment programs with free-market capitalism in the private sector.

If we're serious about the steady state economy for environmental protection, economic sustainability, national security and international stability, we better get serious about maintaining full employment with a mix of public and private sector jobs. FDR was serious about maintaining full employment, even in the midst of *degrowth*, and pulled the US out of the Great Depression. We should be serious about it too, and pull our respective countries away from the depression of Supply Shock, preferably (by far) without the equivalent of a World War II to help "stimulate the economy."

There is at least one other promising approach to the unemployment problem: reducing the time spent working. With this approach, increasing labor productivity is used to give everyone more time off, instead of laying some workers off while others continue working long hours. Unlike the approach of labor intensification, this approach is especially suited to middle- and older-aged workers. Working-time reduction is not only beneficial for maintaining widespread employment in a full-world economy, it helps with achieving that enviable and elusive goal of work-life balance. As with labor intensification, working-time reduction is a practical approach with solid precedents. The Dutch, for example, have demonstrated that working-time reduction and work-life balance can be achieved in a systematic manner with public policies that resonate with the people.[29]

No matter what the approaches to employment, however, none can be successful in the long run unless population is stabilized. Population won't be stabilized without a steady-state paradigm shift. For the steady statesman, putting the horse before the cart means providing leadership in promulgating the steady-state paradigm shift.

The final policy issue from *Enough Is Enough* that warrants attention here is changing the way we measure progress. Actually, we got a start on this subject in Chapter 2, with the stance that GDP itself should not be tampered with. The logic was that GDP is a well-established and quite meaningful indicator of one thing: the size of the economy. It is not GDP itself that needs reform, but rather our interpretation of GDP. Once again, we need to put the horse before the cart. With economic growth at the crossroads, and pursuant to a steady-state paradigm shift, the public and polity will interpret growing GDP as an indication of growing problems, not solutions.

In Chapter 2 we considered the zoological metaphor of an elephant in a cage. Its outgrowing the cage led to a very problematic outcome. It was nothing to encourage, just as growing GDP is nothing to encourage in the age of Supply Shock. Now let's consider a medical metaphor that may lead to a more nuanced understanding of measuring progress.

If you're a doctor with an overweight patient, the last thing you should tell the patient is to throw away the scale. The patient needs that scale now more than ever. It just has to be interpreted in a different light. For example, when the patient was a little kid, it was a good, healthy sign when the scale indicated growth from year to year. When the patient became an adult and reached an optimum weight, that was a good thing too. But now, with an overweight patient, increasing size is a bad thing, and the patient needs to know it.

That's how we should use GDP. GDP is a solid indicator of the economy's size. Sure, economists of yesteryear considered GDP an indicator of welfare, not just of size. To them, a growing GDP was invariably a good thing. They were analogous to a narrow-minded pediatrician with an overweight patient, always prescribing growth. For many decades, they were right, too. But while the patient grew up, many of the neoclassical "doctors" never did, as we saw in Chapter 4.

It's time for them to grow up, but that doesn't mean throwing away their instruments. The doctor with the overweight patient should not take away the scale, but rather *emphasize* it. The patient should monitor that scale closely and, as the readings become larger and larger, become evermore alarmed. But the doctor should also make good use of the blood pressure cuff and the stethoscope, both of which will indicate declining health as the patient balloons into obesity.

Likewise, we ought to supplement GDP with other indicators such as the Genuine Progress Indicator (GPI) and the Happy

Planet Index (HPI). For the global economy and many nations, GPI and HPI will continue to decline as GDP grows beyond optimal levels. With GDP growth now coming at the expense of genuine progress and happiness, we should strive to halt the growth in GDP. That doesn't mean we should stop measuring it; quite the opposite in fact. We'll want to know how we're faring in our progress toward a steady state. GDP will be a key indicator for monitoring such progress.

Of course no metaphor is perfect, and GDP may be even more useful than the medical metaphor suggests. There's a lot of "value added" to GDP monitoring, once we put the horse before the cart. For example, all one needs to *indicate* (as opposed to measure precisely) the loss of biodiversity is GDP.[30] That's because of the fundamental conflict between economic growth and biodiversity conservation, as described in Chapter 8. For the sole purpose of indicating biodiversity loss, there's no need to consider the complex metrics of GPI or HPI.

Now the devil's advocate will ask, "Why not just count the endangered species directly, instead of looking for an indicator like GDP?" The problem is that counting endangered species is akin to counting oil spills. They don't come out and advertise themselves. A spill the size of BP's Deepwater Horizon won't escape notice, nor will the endangerment of a species like the polar bear, but the little spills and the little species are often overlooked and sometimes undetectable. Also, many forces are aligned to prevent the counting and reporting of endangered species, as I learned during my PhD research on the Endangered Species Act.[31] Even if it weren't for these forces, you'd have a hard time monitoring the millions of species on the planet.

This type of problem is why we have indicators to begin with. Although it would be nice to know exactly which species are endangered—and how many barrels of oil are spilled, how low all the aquifers are, how much topsoil is eroded, how many toxins are being emitted, etc.—we cannot know, and even if we could, we probably wouldn't find it worth the expense to ascertain. Not with the trophic theory of money telling us that, to afford such an impossible analysis, we'd have to liquidate the very natural capital we were worried about to begin with (Chapter 7). It is important, however, to have some idea of the magnitude and trends of species endangerment—and oil spillage, aquifer depletion, etc. And it is more than feasible.

Indeed, for many indicators of ecosystem degradation, GDP has at least the following advantages: 1) it is a technically sound indicator, most notably for biodiversity loss and greenhouse gas emissions; 2) it is already calculated with due diligence by governments, saving conservation and environmental organizations the huge amounts of money they would have to spend on more direct measures of environmental impact; 3) GDP data are widely reported by the press.

Finally, there is one thing for which GDP is probably unsurpassed as an indicator. Some may argue that GDP isn't a perfect indicator of greenhouse gas emissions because the carbon intensity of GDP changes. Some may argue that the rate of biodiversity loss changes with the technological regime. Some may argue that, while some water pollutants are increasing as a function of economic growth, others are being phased out. All of them may argue it's no use trying to add up such distinct environmental parameters as climate, biodiversity, air and water in coming up with a broad indicator of environmental protection, because it's like adding apples and oranges.

Yet apples and oranges, along with bacon and bourbon, can all be placed in a basket and weighed. If you ingest a small enough basketful, you'll survive, even if it's all bacon or bourbon. If you ingest a massive pile, it can be all organic apples and you're still doomed. In matters of individual survival and social sustainability alike, size matters.

And so it seems fruitful to recall the definition of economic growth: increasing production and consumption of goods and services *in the aggregate*. GDP is a well-established, consistently calculated measure of economic growth. We also know that there is a fundamental conflict between economic growth and environmental protection. So even if we can't add apples and oranges precisely, we can put two and two together: GDP is clearly an indicator of environmental impact in the aggregate, and may very well be the best such indicator we can hope for.

Some reformers want to dispense of GDP entirely, claiming that it's a meaningless indicator at best and a misleading indicator at worst. Yet clearly it is neither. GDP indicates how much trouble we're getting into — how "obese" we're getting with our global economy — and as an indicator, it cannot be "misleading." Any charge of misleading may only be leveled against mistaken interpreters, such as those who think increasing GDP is a positive sign no matter the historical or ecological context.

The late Donella Meadows once made the excellent point that "we care about what we measure."[32] Some have used this quote to argue for dispensing with GDP and adopting an indicator that reflects what we really care about. Certainly, if we were busy measuring the GPI, for example, we would engender more concern with genuine progress. But Meadows' full point was a little different. She said, "Indicators arise from Values (we measure what we care about) and they create Values (we care about what we measure)." She may as well have said, "First comes the horse; then comes the cart."

That leads to a well developed but simple conclusion, for the metaphor of horse-and-cart has clearly become our underlying theme. Yes, we need public policy reform in order to establish national and global steady state economies at sustainable levels. Furthermore, some may never concede the need for steady state economics without first being able to visualize the policy framework. So yes, the steady statesman must be able to propose and articulate policies such as the Full Seas Act, Natural Resources Cap-and-Trade Act, sectoral salary caps, tax reforms toward stabilizing population, phase-out of fractional reserve banking, and labor-intensive civil services. Yet it should be abundantly clear by now that not a single one of these policies stand a reasonable chance of public dialogue, much less adoption, as long as the overriding policy goal and social mode is economic growth. Attempting to pass any one of these heavy hitters would be a major episode of putting the cart before the horse. That's why successful steady statesmanship—the only kind that matters to the grandkids—requires honest, open, persistent and articulate leadership in raising awareness of the perils of economic growth in the age of Supply Shock.

NOTES

Preface
1. "La Décroissance" literally means "the decline." However, in the context of European politics it refers to the movement to shrink European and global economies for purposes of making them sustainable.

Chapter 1
1. Gleick, *Bottled and Sold*.
2. "According to government and industry estimates, about one fourth of bottled water is bottled tap water (and by some accounts, as much as 40 percent is derived from tap water)—sometimes with additional treatment, sometimes not." Natural Resources Defense Council website, nrdc.org/water/drinking/bw/exesum.asp; accessed July 20, 2012.
3. Glennon, *Unquenchable*.
4. de Villiers, *Water*.
5. Dolan, *Our Poisoned Waters*.
6. Clemens et al., "Technical Concepts Related to Conservation of Irrigation and Rainwater in Agricultural Systems." Authors' calculations pertaining to freshwater withdrawals based on US Geological Survey data.
7. President Franklin Delano Roosevelt made this statement in a 1937 letter to state governors about a potential Uniform Soil Conservation Law. See en.wikiquote.org/wiki/Franklin_D._Roosevelt.
8. Pimentel, *Soil Erosion*.
9. Brown and Wolf (1984).
10. Pimentel, *Soil Erosion*.
11. Etter, "Lofty prices for fertilizer put farmers in a squeeze." *Wall Street Journal*, May 27, 2008.
12. Adelman, *The Genie Out of the Bottle*, 329.
13. Deffeyes, *Hubbert's Peak*. A plethora of information on Peak Oil is now available from numerous sources in academia, business and government, including one of the least political and most objective American agencies, the Government Accountability Office (Cf. GAO, *Crude Oil* in "Literature Cited.").

14. Harris and Bajaj, "As power is restored in India, the 'blame game' over blackouts heats up," *New York Times*, August 1, 2012.
15. The United Nations Food and Agriculture Organization (FAO) estimated that there were 1.38 billion cattle in the world (faostat.fao.org/site/573/default.aspx#ancor) and 398 million acres of paddy rice (faostat.fao.org/site/567/default.aspx#ancor) as of 2009.
16. The annual release of nitrous oxide into the atmosphere due to human economic activity is 7–13 million tons, with nitrogen-based fertilizers typically identified as the primary source (umich.edu/~gs265/society/greenhouse.htm).
17. Klare, *Resource Wars*.
18. There is a burgeoning literature on sea-level rise and the most authoritative publications (with conservative estimates of sea-level rise) are produced by the Intergovernmental Panel on Climate Change. See for example Nicholls et al., "Coastal Systems and Low-Lying Areas."
19. Tidwell, *Bayou Farewell*. Also see the State of Louisiana's website on coastal land loss: coastal.louisiana.gov/index.cfm?md=pagebuilder&tmp=home&pid=112; accessed August 2, 2012.
20. Water withdrawals from coastal aquifers and oil and gas extraction contribute to subsidence of the Louisiana coast, which exacerbates the submersive effects of sea-level rise. Meanwhile the Mississippi River, modified by the Army Corps of Engineers to aid in commercial river traffic, no longer feeds a vast delta with marsh-building sediments, but sends those sediments out to deeper seas like a giant hydraulic conveyor belt. Finally, the Intercoastal Waterway and thousands of miles of oil and gas channels have been cut through the salt marshes, meaning that thousands of miles are now exposed to fetch and wave erosion from all angles. See Tidwell, *Bayou Farewell*.
21. See for example *Global Sea Level Rise Scenarios for the US National Climate Assessment* by Parris et al. (2012).
22. Fox, *Agricide*.
23. EPA, *FY 2002 Annual Report*.
24. See "Dupont complains that Monsanto is running a seed market monopoly," naturalnews.com/028850_DuPont_Monsanto.html#ixzz22Q5Vdgkf; accessed August 2, 2012.
25. Roan, *Ozone Crisis*, 10.
26. The nefarious affair of Big Tobacco's "Seven Dwarves" is documented at senate.ucsf.edu/tobacco/executives1994congress.html; accessed July 21, 2012.
27. Planetary escape should be considered in the context of Ward and Brownlee, *Rare Earth*.

Chapter 2

1. Czech, "The Self-Sufficient Services Fallacy."
2. Another common system of exchange was the "gift economy" (Mauss, *The Gift*). As with bartering, production tended to equal consumption.
3. Value = price × quantity. (See *Deardorffs' Glossary of International Economics*: personal.umich.edu/~alandear/glossary/v.html.) Because inflation is accounted for (i.e., price levels are kept constant) in calculating real GDP, the value of goods and services is a direct function of the quantity thereof. Therefore, GDP is considered a measure of the "amount" (quantity) of goods and services produced.
4. See japan-guide.com/e/e644.html.
5. *CIA World Factbook*, www.cia.gov/library/publications/the-world-factbook/index.html.
6. "GDP dynamics (1969–2009) — 192 countries," www.economicsweb institute.org/ecdata.
7. These figures are reported in 2005 dollars: economicswebinstitute.org/ecdata.
8. These figures were reported in 2012 American dollars by the International Monetary Fund: imf.org/external/pubs/ft/weo/2012/01/weo data/weoselgr.
9. These figures were reported in constant (1992) rupees by the International Monetary Fund: Ibid.
10. The US Environmental Protection Agency reported in 2010 that Americans generated more than two-thirds of a million tons of waste every day, or approximately 250 million tons per year: epa.gov/osw/basic-solid.
11. Korten, *When Corporations Rule the World*.
12. Bakan, *The Corporation*.
13. Figures reported in 2005 dollars: economicswebinstitute.org/ecdata.
14. *CIA World Factbook*: cia.gov/library/publications/the-world-fact book/index.
15. Klare, *Resource Wars*.
16. Daly resigned from the World Bank, disappointed with the bank's destructive pro-growth policies. He became a professor in the School of Public Policy at the University of Maryland, and retired in 2010. He serves on the board of the Center for the Advancement of the Steady State Economy (CASSE) and is CASSE's Economist Emeritus.
17. Perkins, *Confessions of an Economic Hit Man*.
18. Kapsis and Coblentz, *Woody Allen: Interviews*, 64.

19. See google.com/publicdata/explore?ds=d5bncppjof8f9_&met_y=ny_gdp_mktp_cd&tdim=true&dl=en&hl=en&q=global+gdp.
20. See footprintnetwork.org/en/index.php/GFN/page/world_footprint; accessed September 1, 2011.
21. See wikipedia.org/wiki/Uneconomic_growth.
22. Investment "includes business investment in equipments for example and does not include exchanges of existing assets. Examples include construction of a new mine, purchase of software, or purchase of machinery and equipment for a factory. Spending by households (not government) on new houses is also included in Investment. In contrast to its colloquial meaning, 'Investment' in GDP does not mean purchases of financial products. Buying financial products is classed as 'saving,' as opposed to investment. This avoids double-counting: if one buys shares in a company, and the company uses the money received to buy plant, equipment, etc., the amount will be counted toward GDP when the company spends the money on those things; to also count it when one gives it to the company would be to count two times an amount that only corresponds to one group of products. Buying bonds or stocks is a swapping of deeds, a transfer of claims on future production, not directly an expenditure on products." See wikipedia.org/wiki/Gross_domestic_product#Components_of_GDP_by_expenditure.
23. See wikipedia.org/wiki/Index_of_Sustainable_Economic_Welfare.
24. Eisenhower's overriding fear was of inflation, not environmental overshoot. See Collins, *More*.
25. "The Esoteric Emerson" is a fictional title.
26. See degreedirectory.org/articles/What_are_the_Most_Common_College_Majors_in_the_US.
27. Gowdy, *Limited Wants, Unlimited Means*.
28. Economists generally assume that savings equals investment. Therefore, when we say that Consumption = Income − Savings, we are saying that Consumption = Income − Investment, and that Consumption + Investment = Income. As in chapter 2, however, we may consider investment a form of consumption, so that per capita consumption = per capita income.

Chapter 3

1. See "Francois Quesnay" at newworldencyclopedia.org/entry/Fran%C3%A7ois_Quesnay.
2. See wikipedia.org/wiki/Fran%C3%A7ois_Quesnay.
3. Théré and Charles, "The Writing Workshop of Francois Quesnay and the Making of Physiocracy," 3.

4. In addition to Quesnay, prominent physiocrats included Victor de Riquetti (the Marquis de Mirabeau), Anne-Robert-Jacques Turgot and Pierre Samuel du Pont de Nemours. These and other physiocrats were prominent in political developments preceding the French Revolution.
5. Ferguson, *The Ascent of Money*.
6. Quesnay's fear of insulting the king should not be overestimated in retrospect. As Théré and Charles noted, "while Quesnay exercised extreme caution toward the high nobility and the king, he was no fool" (4). After Louis XIV moved the royal residence from Paris to Versailles and established his oppressive control over the lesser nobility (barons, dukes and princes), considerable tension arose between the king and these lower landlords. In other words, Quesnay and the physiocrats could afford to portray the *majority* of landlords (such as the barons, dukes and princes) as "sterile" without offending Louis XV.
7. Some of Quesnay's earliest publications were influenced by Charles-Georges Le Roy, "first lieutenant of the hunting of the park of Versailles" (Théré and Charles, 9).
8. Reill and Wilson, *Encyclopedia of the Enlightenment*, 490.
9. Lasswell, *Politics*.
10. wealthandwant.com/HG/Gems_from_HG.html
11. Galbraith, *Economics in Perspective*, 56.
12. The *Tableau Economique* probably also played a major role. As noted, Quesnay's bust is at the entrance to Thomas Jefferson's Monticello. I also saw the *Tableau* among the books retained in Jefferson's posthumous library at Monticello. Consistent with the *Tableau*, Jefferson thought that agriculture was the bedrock of the budding American economy.
13. Heilbroner, *The Worldly Philosophers*. 49.
14. Smith, *The Wealth of Nations*, 94–95.
15. The phrase "dismal science" was coined by Scottish historian Thomas Carlyle in an 1849 article in *Fraser's Magazine*, "Occasional Discourse on The Nigger Question."
16. Skousen, *The Making of Modern Economics*.
17. Heilbroner, *Worldly Philosophers*, 85.
18. The term "stationary state" was used sparingly by Adam Smith to indicate the cessation of capital accumulation, but Mill expanded the concept into a basic model of political economy.
19. I first encountered this quote of Mill in Daly, "Introduction to Essays Toward a Steady-State Economy," (27–28) and later found the original material in Book IV of Mill's 1848 *Principles of Political Economy*, Chapter VI ("Of the Stationary State").

20. Rostow, *Theorists of Economic Growth*, 117.
21. Heilbroner, *The Worldly Philosophers*, 131.
22. Schneider and Ingram, *Policy Design for Democracy*.
23. Ormerod, *Death of Economics*.

Chapter 4

1. Weintraub, *Neoclassical Economics*. He said simply, "The first to use the term 'neoclassical economics' seems to have been the American economist Thorstein Veblen." The Wikipedia entry for Neoclassical Economics includes: "The term was originally introduced by Thorstein Veblen in 1900, in his *Preconceptions of Economic Science*, to distinguish marginalists in the tradition of Alfred Marshall from those in the Austrian School." wikipedia.org/wiki/Neoclassical_economics; accessed August 13, 2010.
2. For years during the first decade of the 21st century, at least, The New School, a famous progressive university in New York, hosted "The History of Economic Thought," an online encyclopedia. "American Apologists" was designated one of the categories of early 20th-century schools of thought. As of August 2012 the encyclopedia had been taken offline. However, the text of the chapter on the American apologists may found at irfanerdogan.com/dersler/poleconofcom/apologists.htm.
3. Although the "idiot savant" quote is customarily attributed to Leontief, some authors have attributed it to Robert Kuttner. See for example Skousen (2001:96).
4. See Sills and Merton, page 247.
5. Wenzer and West, *The Forgotten Legacy of Henry George*.
6. Gaffney and Harrison, *The Corruption of Economics*, back cover.
7. The importance of the production function in undermining Henry George and the single tax on land is not part of Gaffney's thesis, but it is a logical and probable supplement, as described in Czech, "The Neo-Classical Production Function."
8. See endnote 2.
9. The "marginalist revolution" of 1871–1874 is often cited as the transition from classical to neoclassical economics, as explained in Chapter 4. However, various scholars have shown that the marginalist revolution was actually reconcilable with classical economics and in fact simply formalized a part of Adam Smith's work. See for example Ormerod, *The Death of Economics*, 45–47. In any event, the marginalist revolution is much less relevant to economic growth theory than the transformation of the production function.

10. Ecologists face the same problem in teaching niche theory and use the phrase "n-dimensional hypervolume" to describe niches with more than three dimensions.
11. Czech, "Julian Simon Redux."
12. Gaffney and Harrison, 29.
13. Ibid., 34.
14. Ibid., 36–37.
15. Ibid., 46.
16. Ibid., 47.
17. Ibid., 84.
18. Ibid., 91.
19. Ibid., 117.
20. Knight, "The Fallacies in the Single Tax."
21. Gaffney and Harrison, 118.
22. Plassmann and Tideman, "Frank Knight's Proposal."
23. Skousen, *The Making of Modern Economics*.
24. Whitaker, *The Early Economic Writings of Alfred Marshall, 1867–1890*, 309.
25. Nadeau, *The Wealth of Nature*.
26. Marshall, *Principles of Economics*, xiv.
27. Rostow, *Theorists of Economic Growth*, 188–189.
28. Gaffney and Harrison, *Corruption of Economics*, 65.
29. Keynes, *The General Theory of Employment, Interest and Money*, viii.
30. Ibid., 3.
31. Ibid., 213.
32. The World Bank and International Monetary Fund were created at Bretton Woods, New Hampshire, on July 22, 1944, when representatives of many nations gathered in anticipation of the end of World War II in order to develop a world financial climate more conducive to peace. Keynes was the leading economist in attendance.
33. Daly and Cobb, *For the Common Good*, 209.
34. Of Keynes, Skousen said, "He was not trained in economics, having taken only a single course from Alfred Marshall." *The Making of Modern economics*, 328.
35. Harrod, "An Essay in Dynamic Theory," 261.
36. Ibid., 261.
37. Ibid., 272.
38. Pearce, *The MIT Dictionary of Modern Economics*, 49–50.
39. Ibid., 466.
40. Ibid., 18.

41. Pearce, D. W., "Steady states ignore real life," *The Times Higher Education Supplement*, 2001; timeshighereducation.co.uk/story.asp?storyCode=162901&sectioncode=8.
42. Besomi, "Harrod's Dynamics and the Theory of Growth," 79.

Chapter 5

1. Solow, "A Contribution to the Theory of Economic Growth," 65–94.
2. Various critics will be tempted to jump the gun and adamantly deny that stabilized population, production and consumption would also entail stabilized GDP, perhaps going off into the blogosphere and publicizing their critiques prior to reaching Part 3, which explains why, indeed, stabilized GDP is entailed by stabilized population, production and consumption. Hopefully this endnote will serve as a dose of patience.
3. This was the main point of Frank who encouraged American citizens to save more instead of spending so much on luxury goods and services. This was the patriotic thing to do, Frank said, because it would increase capital investment and therefore economic growth. Frank, *Luxury Fever*.
4. Solow, *Growth Theory*, xxii.
5. See for example, Hobijn, *Identifying Sources of Growth*, 6: "This [capital stock] aggregate is known, due to Solow (1960) as Jelly capital, denoted by J_t..."
6. McCloskey, *The Rhetoric of Economics*, 49.
7. Solow, "The Economics of Resources or the Resources of Economics," 11.
8. Jones, *Introduction to Economic Growth*.
9. Lucas, "Making a Miracle."
10. The concept of increasing returns to scale as the crucial insight of Romer's research is emphasized by Warsh, *Knowledge and the Wealth of Nations*.
11. Jones, *Introduction*, 85. Jones also states in a footnote on page 148: "Without (exogenous) population growth, per capita income growth eventually stops."
12. Theoretically, Romer's model does allow for research productivity to grow over time, even with a constant number of researchers. In the lingo of the model, "$\varphi = 1$." Unfortunately, as Jones explained, the model "blows up" when $\varphi = 1$. In other words, it doesn't work with stabilized population.
13. Luzzati, "Growth Theory and the Environment," 336.
14. The first law of thermodynamics, or the law of conservation of energy, is that energy can be neither created nor destroyed. This law also applies to materials in the sense of Einstein's famous formula, $E = mc^2$.

Matter may be transformed into energy (think of an atomic bomb), or vice versa (particle accelerators transform energy into subatomic particles), but these are transformative and not creative processes. In other words, "You can't get something from nothing." The laws of thermodynamics, vis-à-vis economic growth, are addressed more thoroughly in Chapter 7.
15. A prime example is David Warsh. An economic journalist, in *Knowledge and the Wealth of Nations* he tracked the career of Romer to a level of detail usually reserved for presidents or movie stars. He also praised Romer's endogenous growth theory for its brilliance and denigrated the "doomsayers" with their primitive theory of limits to growth. (It would be interesting to know what Warsh has been thinking since the economic and financial crises of more recent years.)

Chapter 6

1. The ISEE is far from a perfect representative for ecological economics. In its attempts to "reach across the aisle," the ISEE has numerous neoclassical and other pro-growth members. This is an underlying theme of the book—the predominance of neoclassical and other perpetual-growth economics, which permeates even the ecological economics movement. For over a decade, despite the efforts of ISEE members, the ISEE Board of Directors has refused to adopt a position statement on the trade-off between economic growth and environmental protection, even though that trade-off is one of the central concepts of ecological economics.
2. See Turner, "A Comparison of *The Limits to Growth* with Thirty Years of Reality." Turner found that data from 1970–2000 corroborated the "standard run" scenario described in *Limits to Growth*. This scenario reflects "business as usual" with population and economic growth resulting in "overshoot and collapse" in the mid-21st century. Turner also described how earlier critiques of *Limits to Growth* were faulty and based on mischaracterizations that were widely circulated and perpetuated. See also Hall and Day, "Revisiting the Limits to Growth After Peak Oil."
3. One source of this often-cited quote of Einstein is Klein, "Thermodynamics in Einstein's Universe," 509.
4. Although Daly is retired from the University of Maryland, he is on the Board of Directors of the Center for the Advancement of the Steady State Economy (CASSE) and writes a regular column for CASSE's *Daly News*, a blog named after Daly and devoted to steady state economics and policy.

5. *Herman Daly Festschrift* (e-book), *Encyclopedia of the Earth*, National Council for Science and the Environment, eoearth.org/article/Herman_Daly_Festschrift_(e-book)
6. See, for example, Daly, *Ecological Economics and the Ecology of Economics*, 17.
7. Daly, *Ecological Economics*, 60.
8. Lasswell, *Politics*.
9. Norgaard, "Intergenerational Commons, Economism, Globalization, and Unsustainable Development."
10. Czech, "Julian Simon Redux."
11. George Will, opinion columnist with the *Washington Post*, has long been loose with the facts on environmental issues, denying the causes and effects of resource scarcity, pollution and climate change. However, he finds a receptive audience on environmental matters because of the adverse reactions to "doomsday" prophecies. (See steadystate.org/george-will)
12. With the quotation, "Natural resources originate from the mind…," Bradley was describing one of Julian Simon's "energy themes." Clearly he agreed with these themes; the quotation comes from Bradley's acceptance speech of the Julian Simon Award. See heartland.org/policybot/results/10298/Robert_Bradley_receives_Julian_Simon_Award.
13. See senate.ucsf.edu/tobacco/executives1994congress
14. Czech, "American Indians and Wildlife Conservation"; Czech, "Big Game Management on Tribal Lands"; Diamond, *Guns, Germs, and Steel*.
15. Michener, *Poland*.
16. Ferguson, *The Ascent of Money*.
17. McDaniel and Gowdy, *Paradise for Sale*.
18. Czech, "Big Game Management."
19. Diamond, *Guns, Germs, and Steel*. See also, for the North American case especially, Stannard, *American Holocaust*.
20. Geist, *Buffalo Nation*.
21. Oelschlaeger, *Caring for Creation*.
22. Types of energy include kinetic, potential, thermal (heat), chemical, electrical, electrochemical, electromagnetic (light), sound and nuclear.
23. One may hear Einstein's pronouncement at aip.org/history/einstein/voice1.
24. Czech et al., "Economic Associations Among Causes of Species Endangerment in the United States."
25. Odum and Odum, *The Prosperous Way Down*, 67.

26. Less than a year after I wrote chapter 1, a three-part series was published in the 2003 volume of *New Scientist*. The first paragraph summed up the predicament: "Civilization is at a turning point. In the next 50 years, we will experience the biggest surge in energy demand in history. Yet a growing number of experts are warning that the rate at which we are able to pump oil from the ground is likely to peak within a few years. If we want to keep our cars on the roads and the lights on in our homes, we need to find a new source of energy, and fast." newscientist.com/article/mg17924061.000-power-struggle.html; accessed July 21, 2012.

Chapter 7
1. Naisbitt, *Megatrends*.
2. Ferguson, *The Ascent of Money*.
3. Ellison, "What if they held Christmas and nobody shopped?"
4. Fortey (*Life*) provided a rigorous and riveting account of the history of life on earth, describing how biomass and diversity have waxed and waned for four billion years. His account is a tour de force of the division of labor—and limits to growth—in the economy of nature.
5. See energy.saving.nu/biomass/basics; accessed October 12, 2010.
6. Raup and Sepkoski, "Periodic Extinction of Families and Genera."
7. In more technical terms, the second law of thermodynamics states that entropy constantly increases in a closed system, meaning that energy is constantly more diffused. Given $E = mc^2$, the second law implies that materials disintegrate, an implication that Nicholas Georgescu-Roegen proposed as a fourth law of thermodynamics.
8. Roberts, "Academy Panel Split on Greenhouse Adaptation."
9. Daly, "When Smart People Make Dumb Mistakes."
10. Schelling, "The Cost of Combating Global Warming," 9.
11. I described the precepts of the trophic theory of money in Chapter 3 of *Shoveling Fuel for a Runaway Train*, (Czech 2000), but did not coin the phrase "trophic theory of money" therein. See also "The Trophic Structure of the Economy" at steadystate.org/wp-content/uploads/2009/12/CASSE_Brief_TrophicStructureOfTheEconomy.pdf.
12. Skidelsky, *Keynes*, 7.
13. Ederer, *The Evolution of Money*.
14. Ward and Brownlee, *Rare Earth*.
15. The fact that there is tremendous political pressure to demonstrate GDP growth should be a matter of political common sense. For some of the ways this pressure manifests in government affairs and

international development, see Perkins, *Confessions of an Economic Hit Man*.
16. See, for example, www.dieoff.org.

Chapter 8
1. Endangered Species Act, 16 U.S.C. sections 1531(a)(1).
2. Czech, "Prospects for reconciling the conflict between economic growth and biodiversity conservation with technological progress."
3. Czech et al., "Economic Associations Among Causes of Species Endangerment in the United States."
4. Yaffee, *The Wisdom of the Spotted Owl*.
5. Kurlansky, *Cod*.
6. By August 20, 2012, at the time of this writing, approximately 39 million cars had been produced in 2012 and the schedule for 2012 production was set to exceed 60 million: worldometers.info/cars.
7. See worldometers.info/cars.
8. Heiskanen and Jalas, *Dematerialization Through Services*.
9. Begun and Eicher, *In Search of a Sulphur Dioxide Environmental Kuznets Curve*.
10. Stern, "The Rise and Fall of the Environmental Kuznets Curve."
11. Czech et al., "Economic Associations," 593.
12. Asafu-Adjaye, "Biodiversity Loss and Economic Growth"; Naidoo and Adamowicz, "Effects of Economic Prosperity on Numbers of Threatened Species"; see also Clausen and York, "Global Biodiversity Decline of Marine and Freshwater Fish" regarding fish.
13. Dietz et al., "Driving the Human Ecological Footprint," 17.
14. Ibid., 16.
15. Perelman, "R&D, Technological Progress and Efficiency Change in Industrial Activities."
16. Fried et al., *The Measurement of Productive Efficiency: Techniques and Applications*.
17. Li et al., "Product Innovation and Process Innovation in SOEs."
18. Wils, "End-Use or Extraction Efficiency in Natural Resource Utilization."
19. Dietz et al., "Driving the Human Ecological Footprint."
20. Maddison, "Growth and Slowdown in Advanced Capitalist Economies," 649.
21. Sharpe, "Ten Productivity Puzzles Facing Researchers."
22. Oguchi, "Productivity Trends in Asia since 1980," 69.
23. Sodhi et al., "Southeast Asian Biodiversity," 654.

24. Some technological progress also results from "learning by doing" (i.e., on-the-job innovation), but with diminishing returns and temporarily for a given technology (Lucas, *Making a Miracle*, 251).
25. Duga and Stadt, "Global R&D Report."
26. National Science Foundation, *National Patterns of R&D Resources*.
27. Ibid.
28. Anderson and Cavanagh, *Top 200*.
29. Duga and Stadt, "Global R&D Report."
30. Ruttan, *Technology, Growth, and Development*, 75.
31. National Science Foundation, *Academic Research and Development Expenditures*.
32. Duga and Stadt, "Global R&D Report."
33. American Association for the Advancement of Science, *Government R&D Expenditures*.
34. Duga and Stadt, "Global R&D Report."
35. National Science Foundation, *National Patterns of R&D Resources*.
36. Sinclair et al., "An In-Depth Study of America's Largest Non-Profits."
37. Hecker, "Occupational Employment Projections to 2010."
38. Brown et al., "Measuring Research and Development Expenditures in the U.S. Economy," 2.
39. Winthrop et al., "Government R&D Expenditures."
40. Ruttan, *Technology, Growth, and Development*, 87 (italics in original).
41. Pearce, *The MIT Dictionary of Modern Economics*, 122.
42. Denison, *Accounting for United States Economic Growth: 1929–1969*.
43. Rostow, *Theorists of Economic Growth from David Hume to the Present*, 371.
44. Gordon, *An Empire of Wealth*.
45. Kingdon, *Self-Made Man*.

Chapter 9
1. Collins, *More*.
2. Ransom, "Reconstructed but Unregenerate," 8.
3. The common form of political economy prior to the industrial revolution and the era of capitalism was called mercantilism, which established that the primary economic strategy of a nation-state was to accumulate bullion for the sake of consolidating military might.
4. Roosevelt, *Public Papers and Addresses*, 1:743, 750–752.
5. Collins, *More*, 11.
6. Ibid., 8.
7. Ibid., 13.

8. NSC-68: *United States Objectives and Programs for National Security*.
9. See digitalhistory.uh.edu/disp_textbook.cfm?smtID=3&psid=3630
10. 15 U.S.C. § 1021.
11. Collins, *More*, 20.
12. Ibid., 21.
13. Ibid., 22.
14. For example, during the 1996 campaign, vice presidential candidate Jack Kemp exclaimed, "We should *double* the rate of growth, and we should *double* the size of the American economy!" (As I noted in *Shoveling Fuel for a Runaway Train*, there's never been a redder face under whiter hair.) Al Gore responded with the impeccably wry retort, "Well, the economy is growing very strongly right now.... The average growth rate is also coming up. It is higher than in either of the last two *Republican* administrations."
15. Collins, *More*, 25.
16. See especially Perkins, *Confessions of an Economic Hit Man*.
17. Collins, *More*, 37.
18. Ibid., 43.
19. Ibid., 42.
20. Ibid., 46.
21. Tobin, "Economic Growth as an Objective of Government Policy," 1.
22. Tobin, *The New Economics One Decade Older*, 3.
23. Haveman, R. H., ed., *A Decade of Federal Antipoverty Programs*, 11.
24. Collins, *More*, 60.
25. Ibid., 99.
26. Ibid., 68.
27. Ibid., 99.
28. Ibid., 134, indicates that President Carter read Schumacher's book.
29. Meadows et al., *Limits to Growth*.
30. Inaugural Addresses of the Presidents of the United States: Jimmy Carter, January 20, 1977: bartleby.com/124/pres60
31. Deregulation of commercial air travel was one of the most notorious "reforms" of Reaganomics. It caused tremendous turmoil in the airline industry, monopolization, and higher prices for travelers. See Petzinger, *Hard Landing*.
32. Collins, *More*, 203.
33. It is "nearly" impossible only because Gore wrote: "The current debate over sustainable development is based on the widespread recognition that many investments by major financial institutions, such as the World Bank, have stimulated economic development in the

Third World by encouraging the short-term exploitation of natural resources, thus emphasizing short-term cash flow at the expense of longer-term, sustainable growth." (*Earth in the Balance*, 191.) Did Gore consciously decide to participate in the rhetoric of "sustainable growth" despite recognizing (as he must have, based on the rest of *Earth in the Balance*) that there is a limit to economic growth in the conventional sense of the term (i.e., increasing production and consumption of goods and services in the aggregate)? We may never know.

34. For international readers unaware of the significance of the 2000 presidential election, it was the most contested presidential race in American history. Ultimately the Supreme Court had to decide who would be president and selected the anti-environmental George W. Bush. Ralph Nader of the Green Party was considered the "spoiler," grabbing enough of the Democratic vote to turn the election in Bush's favor. The significance of the Earth Day 2000 Gore speech cannot be proven, but it is likely that many who were present decided to vote for Nader (and LaDuke) instead of Gore, and convinced others (including in other states) to do likewise. Their verdict was that, when it came to the perpetual pandering to economic growth interests, Democrats were not significantly different from Republicans. A vote for the Green Party was a strong message to both parties that washed-out environmental rhetoric was no longer acceptable.
35. Czech et al., "The Iron Triangle."
36. Public Papers of the Presidents, Dwight D. Eisenhower, 1960, 1035–1040; may be found at coursesa.matrix.msu.edu/~hst306/documents/indust
37. Leonhardt, "Obamanomics."
38. The position on economic growth is available at steadystate.org/sign-the-position.

Chapter 10

1. Supreme Court Justice Potter Stewart, *Jacobellis v. Ohio*, 1964.
2. See cbsnews.com/stories/2006/01/13/opinion/meyer/main1206612.
3. No ad hominem attack is attempted or implied here, because there is no argument with anything said by Schwarzman. Schwarzman's consumption behavior is simply an illustration of (not an argument for) conspicuous consumption.
4. Gross, "The Golden Ass."
5. See wikipedia.org/wiki/740_Park_Avenue.
6. Kantrow, "Media Maneuvers."

7. Bartiromo and Whitney, *The Weekend That Changed Wall Street*.
8. Patterson, *The Quants*, 158.
9. de la Merced, "Inside Stephen Schwarzman's Birthday Bash."
10. Ibid.
11. See today.msnbc.msn.com/id/38994242/ns/today-books.
12. See wallstreetoasis.com/forums/4-million-birthday-party-steve-schwarzman-of-blackstone.
13. Kantrow, "Media Maneuvers."
14. M. Wackernagel et al., "The Ecological Footprint of Cities and Regions."
15. Leopold, *A Sand County Almanac*, ix. Ecologists would also be interested to note that this was one of the last sentences Leopold wrote for publication prior to his untimely death in 1948. It appears near the end of the foreword to *A Sand County Almanac*, Leopold's masterpiece, calling for a "land ethic." This chronology suggests that Leopold was gravitating toward ecological economics as the answer to environmental deterioration.

Chapter 11

1. Wen Jiabao's complete sentence was, "We absolutely cannot again sacrifice the environment as the cost for high-speed growth, to have blind development, and in that way to create over-capacity and put greater pressure on the environment and resources." See guardian.co.uk/environment/2011/feb/28/china-gdp-emissions.
2. These words were written approximately three months prior to the March 11, 2011 earthquake and tsunami that caused one of the worst nuclear disasters in history at the Fukushima plant in Japan.
3. Schneider and Ingram, *Policy Design for Democracy*.
4. Rolston, "Life in Jeopardy on Private Property," 43.
5. 15 U.S.C. 58. Full Employment and Balanced Growth Act of 1978.
6. Ormerod, *The Death of Economics*.
7. Czech. *The Endangered Species Act*.
8. Czech and Krausman, 2001.
9. Yaffee, *Prohibitive Policy*, 13.
10. Reffalt, "The Endangered Species List," 78.
11. See *A Legislative History of the Endangered Species Act of 1973*, as amended in 1976, 1977, 1978, 1979 and 1980, together with a section by section index prepared by the Congressional Research Service of the Library of Congress for the Committee on Environment and Public Works US Senate, ftp.resource.org/courts.gov/juris/j2262_01.sgml.
12. See geography.about.com/cs/countries/a/numbercountries.

13. The hundred nations with the lowest per capita GDP, in alphabetical order, are: Afghanistan, Albania, Algeria, Angola, Armenia, Bangladesh, Belize, Benin, Bhutan, Bolivia, Bosnia and Herzegovina, Burkina Faso, Burma, Burundi, Cambodia, Cameroon, Cape Verde, Central African Republic, Chad, Comoros, Democratic Republic of the Congo, Republic of the Congo, Côte d'Ivoire, Djibouti, Dominica, Dominican Republic, Ecuador, Egypt, El Salvador, Eritrea, Ethiopia, Fiji, The Gambia, Georgia, Ghana, Guatemala, Guinea, Guinea-Bissau, Guyana, Haiti, Honduras, India, Indonesia, Iran, Iraq, Jamaica, Jordan, Kenya, Kiribati, Kosovo, Kyrgyzstan, Laos, Lesotho, Liberia, Republic of Macedonia, Madagascar, Malawi, Mali, Mauritania, Moldova, Mongolia, Morocco, Mozambique, Nepal, Nicaragua, Niger, Nigeria, Pakistan, Papua New Guinea, Paraguay, Peru, Philippines, Rwanda, Samoa, São Tomé and Príncipe, Senegal, Sierra Leone, Solomon Islands, Sri Lanka, Sudan, Swaziland, Syria, Tajikistan, Tanzania, Thailand, Timor-Leste, Togo, Tonga, Tunisia, Turkmenistan, Tuvalu, Uganda, Ukraine, Uzbekistan, Vanuatu, Vietnam, Yemen, Zambia, Zimbabwe.
14. The "G10" comprise the ten nations highest in per capita GDP according to the 2010 *CIA Factbook*. See wikipedia.org/wiki/List_of_countries_by_GDP_(PPP)_per_capita.
15. Dietz and O'Neill, *Enough Is Enough*.
16. The US military is often viewed as a global police force and is a major petroleum consumer. See www.energybulletin.net/stories/2006-02-26/us-military-oil-consumption.
17. Sorkin, *Too Big to Fail*, 43.
18. Daley, J., "Paulson plans to donate £410m fortune to environmental causes," *The Independent*, January 16, 2007.
19. Heilprin, J., "A Global Warming Believer in Bush Cabinet," libertypost.org/cgi-bin/readart.cgi?ArtNum=144182, June 2, 2006.
20. Sorkin, *Too Big to Fail*, 43.
21. Heilprin, J., "A Global Warming Believer in Bush Cabinet."
22. See minyanville.com/dailyfeed/what-will-fed-heads-do.
23. See brainyquote.com/quotes/quotes/h/henryford136294.
24. See uhuh.com/unreal/lincoln.htm. Lincoln's comment followed the passage of the National Banking Act of 1863, the forerunner of the Federal Reserve Act of 1913. With context included, Lincoln said, "I see in the near future a crisis approaching that unnerves me and causes me to tremble for the safety of my country. Corporations have been enthroned, an era of corruption in high places will follow, and the money

power of the country will endeavor to prolong its reign by working upon the prejudices of the people, until the wealth of the nation is aggregated in a few hands, and the Republic is destroyed."

25. See brainyquote.com/quotes/quotes/h/henryford136294.
26. See wikipedia.org/wiki/Reserve_requirement.
27. See wikipedia.org/wiki/List_of_community_currencies_in_the_United_States.
28. The healthy nature of reasonable but rigorous outdoor labor should resonate with common sense. For me this is no mere hypothesis, having worked for the latter-day equivalent of the CCC, the Youth Conservation Corps (YCC), as well as in other strenuous outdoor capacities for some years before and after.
29. Schor, "Sustainable Consumption and Worktime Reduction."
30. Czech et al. (2005).
31. Czech and Krausman (2001).
32. Meadows, "Indications and Information Systems for Sustainable Development," viii.

LITERATURE CITED

American Association for the Advancement of Science, *Government R&D Expenditures by Country and Socioeconomic Objective, 1999*, AAAS, 2002.

Adelman, M. A., *The Genie Out of the Bottle: World Oil Since 1970*, MIT Press, 1995.

Anderson, Sarah and John Cavanagh, *Top 200: The Rise of Corporate Global Power*, Institute for Policy Studies, 2000.

Asafu-Adjaye, John, "Biodiversity Loss and Economic Growth: A Cross-Country Analysis," *Contemporary Economic Policy*, 21, no. 2 (2003): 173–185.

Ausubel, Jesse H., "Can Technology Spare the Earth?" *American Scientist*, 84, no. 2, (1996): 166–178.

Bakan, Joel, *The Corporation: The Pathological Pursuit of Profit and Power*, Free Press, 2004.

Bartiromo, Maria with Catherine Whitney, *The Weekend that Changed Wall Street: An Eyewitness Account*, Portfolio/Penguin, 2010.

Beder, Sharon, *Global Spin: The Corporate Assault on Environmentalism* (rev. ed.) Chelsea Green, 2002.

Begun, Jeffrey and Theo S. Eicher. *In Search of a Sulphur Dioxide Environmental Kuznets Curve: A Bayesian Model Averaging Approach*, Working Paper no. 79, Center for Statistics and the Social Sciences, University of Washington, Seattle, 2007.

Besomi, Daniele, "Harrod's dynamics and the theory of growth: The story of a mistaken attribution," *Cambridge Journal of Economics*, 25, no. 1, (2001): 79–96.

Brown, Lawrence D., Thomas J. Plewes and Marissa A. Gerstein, editors, *Measuring Research and Development Expenditures in the U.S. Economy*, National Academies Press, 2004.

Brown, Lester R. and Edward C. Wolf, *Soil Erosion: Quiet Crisis in the World Economy*, Worldwatch Paper 60, Worldwatch Institute, 1984.

Clausen, Rebecca and Richard York, "Global biodiversity decline of marine and freshwater fish: A cross national analysis of economic, demographic, and ecological influences," *Social Science Research*, 37, no. 4, (2008): 1310–1320.

Clemens, A.J., R.G. Allen and C.M. Burt, "Technical concepts related to conservation of irrigation and rainwater in agricultural systems," *Water Resources Research*, 44, W00E03, 2008.

Clinton, William J., *Public Papers of the Presidents of the United States, Presidential Documents, January 1 to June 26, 2000*, National Archives and Records Administration, 2001.

Collins, Robert M., *More: The Political Economy of Growth in Postwar America*, Oxford University Press, 2000.

Costanza, Robert, Ralph d'Arge, Rudolf de Groot et al, "The value of the world's ecosystem services and natural capital," *Nature*, 387 (1997): 253–260.

Cramer, Gail and Clarence W. Jensen, *Agricultural Economics and Agribusiness*, 6th ed., John Wiley and Sons, 1994.

Czech, Brian, "American Indians and wildlife conservation," *Wildlife Society Bulletin*, 23, no. 4, (1995): 568–573.

———, "Big Game Management on Tribal Lands," in Stephen Demarais and Paul R. Krausman, eds., *Ecology and Management of Large Mammals in North America*, 277–289, Prentice Hall, 1999.

———, *The Endangered Species Act, American Democracy, and an Omnibus Role for Public Policy*, PhD dissertation, University of Arizona, Tucson, 1997.

———, *Shoveling Fuel For A Runaway Train: Errant Economists, Shameful Spenders, And A Plan To Stop Them All*, University of California Press, 2000.

———, "Julian Simon Redux," *Conservation Biology*, 16 (2002): 570–571.

———, "Technological Progress and Biodiversity Conservation: A Dollar Spent a Dollar Burned," *Conservation Biology*, 17, no. 5 (2003), 1455–1457.

———, "Prospects for reconciling the conflict between economic growth and biodiversity conservation with technological progress," *Conservation Biology*, 22, no. 6 (2008), pp. 1389–1398.

———, "The neoclassical production function as a relic of anti-George politics: Implications for ecological economics," *Ecological Economics*, 68 (2009): 2193–2197.

———, "The self-sufficient services fallacy," *Frontiers in Ecology and the Environment*, 7, no. 5 (2009): 240–241.

Czech, Brian, Eugene Allen, David Batker et al, "The iron triangle: why The Wildlife Society needs to take a position on economic growth," *Wildlife Society Bulletin*, 31, no. 2, (2003): 574–577.

Czech, B. and Paul R. Krausman, "Distribution and Causation of Species Endangerment in the United States," *Science*, 277, (1997): 1116–1117.

———, "Public Opinion on Endangered Species Conservation and Policy," *Society and Natural Resources*, 12 (1999): 469–479.

———, *The Endangered Species Act: History, Conservation Biology, and Public Policy*, Johns Hopkins University Press, 2001.

Czech, Brian, Paul R. Krausman and Patrick K. Devers, "Economic Associations Among Causes of Species Endangerment in the United States," *BioScience*, 50, no. 7, (2000): 593–601.

Czech, B., D. L. Trauger, J. Farley, R. Costanza, H. E. Daly, C. A. S. Hall, R. F. Noss, L. Krall and P. R. Krausman, "Establishing Indicators for Biodiversity," *Science*, 308 (2005): 791–792.

Daly, Herman E., "Introduction to *Essays toward a Steady-State Economy*, in Herman E. Daly and Kenneth N. Townsend, eds., *Valuing the Earth: Economics, Ecology, Ethics*, 11–47, MIT Press, 1993.

———, *Beyond Growth: The Economics of Sustainable Development*, Beacon Press, 1997.

———, *Ecological Economics and the Ecology of Economics: Essays in Criticism*, Edward Elgar, 1999.

———, "When smart people make dumb mistakes," *Ecological Economics*, 34, (2000): 1–3.

Daly, Herman E. and John B. Cobb Jr., *For the Common Good: Redirecting the Economy toward Community, the Environment, and a Sustainable Future*, Beacon Press, 1994.

Daly, Herman E. and Kenneth N. Townsend, eds., *Valuing the Earth: Economics, Ecology, Ethics*, MIT Press, 1993.

Deffeyes, Kenneth S., *Hubbert's Peak: The Impending World Oil Shortage*, Princeton University Press, 2001.

de la Merced, Michael J., *Inside Stephen Schwarzman's Birthday Bash*, New York Times Dealbook, February 14, 2007; archived at dealbook.nytimes.com/2007/02/14/inside-stephen-schwarzmans-birthday-bash/

Denison, Edwards F., *Accounting for United States Economic Growth: 1929–1969*, Brookings Institution, 1974.

de Villiers, Marq, *Water: The Fate of Our Most Precious Resource*, Houghton Mifflin, 2001.

Diamond, Jared M., *Guns, Germs, and Steel: The Fates of Human Societies*, W. W. Norton, 1997.

Dietz, Rob and Dan O'Neill, *Enough Is Enough: Building a Sustainable Economy in a World of Finite Resources*, Berrett-Koehler, 2013.

Dietz, Thomas, Eugene A. Rosa and Richard York. "Driving the human ecological footprint," *Frontiers in Ecology and the Environment*, 5, no. 1 (2007): 13–18.

Dolan, Edward F., *Our Poisoned Waters*, Cobblehill Books, 1997.

Duga, J. and T. Stadt, "Global R&D report: changes in the R&D community," *R&D Magazine*, (September 2005): 1–17.

Ederer, Rupert J., *The Evolution of Money*, Public Affairs Press, 1964.

Ellison, Katherine, "What if they held Christmas and nobody shopped?" *Frontiers in Ecology and the Environment*, 10, no. 6 (2008): 568.

Environmental Protection Agency, *FY 2002 Annual Report*, Office of Pesticide Programs, US Environmental Protection Agency, 2002.

———, *Municipal Solid Waste Generation, Recycling, and Disposal in the United States: Facts and Figures for 2010*, US Environmental Protection Agency, 2010; archived at: epa.gov/wastes/nonhaz/municipal/msw99.htm

Etter, Lauren, "Lofty Prices for Fertilizer Put Farmers in a Squeeze," *Wall Street Journal*, May 27, 2008.

Ferguson, Niall, *The Ascent of Money: A Financial History of the World*, Penguin, 2008.

Fortey, Richard, *Life: A Natural History of the First Four Billion Years of Life on Earth*, Alfred A. Knopf, 1998.

Fox, Michael W., *Agricide: The Hidden Crisis That Affects Us All*, 2nd ed., Krieger Publishing, 1996.

Frank, Robert H., *Luxury Fever: Money and Happiness in an Era of Excess*, Princeton University Press, 1999.

Fried, Harold O., C. A. Knox Lovell and Shelton S. Schmidt, eds., *The Measurement of Productive Efficiency: Techniques and Applications*, Oxford University Press, 1993.

Gaffney, Mason and Fred Harrison, *The Corruption of Economics*, Shepheard-Walwyn, 1994.

Galbraith, John Kenneth, *Economics in Perspective: A Critical History*, Houghton Mifflin, 1987.

Geist, Valerius, *Buffalo Nation: History and Legend of the North American Bison*, Voyageur Press, 1996.

Georgescu-Roegen, Nicholas, *The Entropy Law and the Economic Process*, Harvard University Press, 1971.

Gleick, Peter H., *Bottled and Sold: The Story Behind our Obsession with Bottled Water*, Island Press, 2010.

Glennon, Robert, *Unquenchable: America's Water Crisis and What To Do About It*, Island Press, 2009.

Gordon, John S., *An Empire of Wealth: The Epic History of American Economic Power*, Harper Collins, 2004.

Gore, Al. *Earth in the Balance: Ecology and the Human Spirit*, Rodale Books, 2006.

Government Accountability Office, *Crude Oil: Uncertainty about Future Oil Supply Makes It Important to Develop a Strategy for Addressing a Peak and Decline in Oil Production*, GAO-07-283, Government Accountability Office, 2007.

Gowdy, John, ed., *Limited Wants, Unlimited Means: A Reader on Hunter-Gatherer Economics and the Environment*, Island Press, 1997.

Gross, Daniel, "The Golden Ass: How Blackstone CEO Steve Schwarzman's antics may cost him and his colleagues billions of dollars," *Slate*, 2007; archived at: slate.com/articles/business/money box/2007/06/the_golden_ass.html

Hall, Charles and John Day, "Revisiting the Limits to Growth after Peak Oil" *American Scientist*, 97, (2009): 230–238.

Harris, Gardiner and Vikas Bajaj, "As Power Is Restored in India, the 'Blame Game' Over Blackouts Heats Up," *New York Times*, August 1, 2012.

Harrod, Roy, "An Essay in Dynamic Theory," *Economic Journal*, 49, (1939): 14–33.

Haveman, R. H., ed., *A Decade of Federal Antipoverty Programs: Achievements, Failures, and Lessons*, Academic Press, 1977.

Hecker, Daniel, "Occupational employment projections to 2010," *Monthly Labor Review*, Bureau of Labor Statistics, November 2001, 57–84.

Heilbroner, Robert L., *The Worldly Philosophers: The Lives, Times, and Ideas of the Great Economic Thinkers*, 6th ed., Simon and Schuster, 1992.

Heiskanen, Eva, and Mikko Jalas, *Dematerialization Through Services — A Review and Evaluation of the Debate*, Finnish Ministry of the Environment, 2000; archived at ymparisto.fi/default.asp?contentid=73155 &lan=en

Hobijn, B., *Identifying Sources of Growth*, Domestic Research Division, Federal Reserve Bank of New York, 2000.

Jones, Charles, *Introduction to Economic Growth*, W. W. Norton, 1998.

Kantrow, Yvette, "Media Maneuvers: The New Yorker Profiles Stephen Schwarzman, aka the 8 Billion Dollar Man," *The Huffington Post*, February 8, 2008; huffingtonpost.com/yvette-kantrow/media-manuvers-the-new-yo_b_85775.html

Kapsis, Robert and Kathie Coblentz, eds., *Woody Allen: Interviews*, University Press of Mississippi, 2006.

Kates, R. W., W. C. Clark, R. Corell et al, "Sustainability Science," *Science*, 292, (2001): 641–642.

Keynes, John Maynard, *The General Theory of Employment, Interest and Money*, Harcourt and Brace, 1936.

Kingdon, Jonathan, *Self-Made Man: Human Evolution from Eden to Extinction?* John Wiley and Sons, 1993.

Klare, Michael T., *Resource Wars: The New Landscape of Global Conflict*, Holt Paperbacks, 2002.

Klein, M. J., "Thermodynamics in Einstein's Universe," *Science*, 157, (1967): 509–516.

Knight, Frank H., "The Fallacies in the 'Single Tax,'" *The Freeman*, 3, no. 23, (1953): 809–811.

Korten, David C., *When Corporations Rule the World*, 2nd ed., Berrett-Koehler, 2001.

Krishnan, Rajaram, Jonathan Harris and Neva Goodwin, eds., *A Survey of Ecological Economics*, Island Press, 1995.

Kurlansky, Mark, *Cod: A Biography of the Fish that Changed the World*, Walker, 1997.

Lasswell, Harold D., *Politics: Who Gets What, When, and How?* McGraw-Hill, 1936.

Leonhardt, David, "Obamanomics," *New York Times Magazine*, August 24, 2008.

Leopold, Aldo, *A Sand County Almanac: And Sketches Here and There*, Oxford University Press, 1949.

Li, Yuan, Yi Liu and Feng Ren, "Product innovation and process innovation in SOEs: evidence from the Chinese transition," *Journal of Technology Transfer*, 32, no.1 (2007): 63–85.

Lucas, Robert E. Jr., "Making a Miracle," *Econometrica*, 61, no. 2 (1993) 251–272.

Luzzati, Tommaso, "Growth theory and the environment: how to include matter without making it really matter," in Neri Salvadori, ed., *The Theory of Economic Growth: A 'Classical' Perspective*, 329–341, Edward Elgar, 2003.

Maddison, Angus, "Growth and Slowdown in Advanced Capitalist Economies: Techniques of Quantitative Assessment," *Journal of Economic Literature*, 25 (1987): 649–698.

Marshall, Alfred, *Principles of Economics*, 8th ed., MacMillan, 1930.

Mauss, Marcel, *The Gift: The Form and Reason for Exchange in Archaic Societies*, W. W. Norton and Company, 2000.

McCloskey, Deirdre N., *The Rhetoric of Economics*, 2nd ed., University of Wisconsin Press, 1998.

McDaniel, Carl and John M. Gowdy, *Paradise for Sale: A Parable of Nature*, University of California Press, 2000.

McNeill, J. R., *Something New Under the Sun: An Environmental History of the Twentieth-Century World*, W. W. Norton, 2000.

Meadows, Donella, *Indicators and Information Systems for Sustainable Development: A Report to the Balatan Group*, The Sustainability Institute, 1998.

Meadows, Donella H., Dennis L. Meadows and Jorgen Randers, *The Limits to Growth: A Report for the Club of Rome's Project on the Predicament of Mankind*, Universe Books, 1972.

Michener, James A., *Poland*, Fawcett, 1984.

Mill, J. S., *Principles of Political Economy, With Some of Their Applications to Social Philosophy*, rev. ed., Colonial Press, 1900.

Nadeau, Robert, *The Wealth of Nature: How Mainstream Economics Has Failed the Environment*, Columbia University Press, 2003.

Naidoo, Robin and Wiktor L. Adamowicz, "Effects of Economic Prosperity on Numbers of Threatened Species," *Conservation Biology*, 15, no. 4, (2001): 1021–1029.

Naisbitt, John, *Megatrends: Ten New Directions Transforming Our Lives*, Warner Books, 1982.

National Science Foundation, *Academic Research and Development Expenditures: Fiscal Year 2006*, Division of Science Resources Statistics, National Science Foundation, 2007.

———, *National Patterns of R&D Resources: 2006 Data Update*, Division of Science Resources Statistics, National Science Foundation, 2007.

Nelson, Richard R., *The Sources of Economic Growth*, Harvard University Press, 1996.

Nicholls, Robert J., Poh Poh Wong et al, "Coastal systems and low-lying areas," in M. L. Parry et al., eds, *Climate Change 2007: Impacts, Adaptation and Vulnerability*, 315–356, Cambridge University Press, 2007.

Norgaard, Richard B., "Intergenerational Commons, Economism, Globalization, and Unsustainable Development," *Advances in Human Ecology*, 4 (1995): 141–171.

Odum, Howard T. and Elisabeth C. Odum, *A Prosperous Way Down: Principles and Policies*, University Press of Colorado, 2001.

Oelschlaeger, Max, *Caring for Creation: An Ecumenical Approach to the Environmental Crisis*, Yale University Press, 1994.

Oguchi, Noriyoshi, "Productivity Trends in Asia Since 1980," *International Productivity Monitor*, 10, no. 1 (2005): 69–78.

Ormerod, Paul, *The Death of Economics*, John Wiley and Sons, 1997.

Patterson, Scott, *The Quants: How a New Breed of Math Whizzes Conquered Wall Street and Nearly Destroyed It*, Crown Business, 2010.

Parris, A., P. Bromirski, V. Burkett, D. Cayan, M. Culver, J. Hall, R. Horton, K. Knuuti, R. Moss, J. Obeysekera, A. Sallenger, and J. Weiss. 2012.

Global Sea Level Rise Scenarios for the US National Climate Assessment. NOAA Technical Memo OAR CPO-1. 37 pp.

Pearce, David W., *The MIT Dictionary of Modern Economics*, 4th ed., MIT Press, 1992.

Perelman, Sergio, "R&D, Technological Progress and Efficiency Change in Industrial Activities." *Review of Income and Wealth*, 41, no. 3 (1995): 349–366.

Perkins, John, *Confessions of an Economic Hit Man*, Plume, 2005.

Petzinger, Thomas Jr., *Hard Landing: The Epic Contest for Power and Profits That Plunged the Airlines Into Chaos*, Times Business, 1995.

Pew Research Center. *People and the Press: 1999 Values Update Survey*, Pew Research Center, 1999.

Pimentel, David, "Soil Erosion: A Food and Environmental Threat," *Environment, Development and Sustainability*, 8 (2006): 119–137.

Plassman, Florenz and T. Nicolaus Tideman, "Frank Knight's Proposal to End Distinctions Among Factors of Production and His Objection to the Single Tax," *History of Political Economy*, 36, no. 3 (2004): 505–519.

Ransom, John Crowe, "Reconstructed but Unregenerate," in Twelve Southerners, *I'll Take My Stand: The South and the Agrarian Tradition*, 75th Anniversary ed., Louisiana State University Press, 2006.

Raup, David and J. John Sepkoski, "Periodic Extinction of Families and Genera," *Science*, 231, no. 4740, (1986): 833–836.

Reffalt, William, "The Endangered Species Lists: Chronicles of Extinction?" in Kathryn A. Kohm, ed. *Balancing on the Brink of Extinction: The Endangered Species Act and Lessons for the Future*, 77–85, Island Press, 1991.

Reill, Peter Hans and Ellen Judy Wilson, *Encyclopedia of the Enlightenment*, Facts on File, 2004.

Roan, Sharon, *Ozone Crisis: The 15-Year Evolution of a Sudden Global Emergency*, John Wiley and Sons, 1989.

Roberts Leslie, "Academy Panel Split on Greenhouse Adaptation," *Science*, 253, no. 5025, (1991): 1206.

Rolston, Holmes III, "Life in Jeopardy on Private Property," in Kathryn A. Kohm, ed., *Balancing on the Brink of Extinction: The Endangered Species Act and Lessons for the Future*, 43–61, Island Press, 1991.

Romer, Paul M., "Endogenous Technological Change," *Journal of Political Economy*, 98, no. 5, (1990): S71–S102.

Roosevelt, Franklin D., *The Public Papers and Addresses of Franklin D. Roosevelt, Volume 1*, ed., S. Rosenman, Random House, 1938.

Rostow, W. W., *Theorists of Economic Growth from David Hume to the Present: With a Perspective on the Next Century*, Oxford University Press, 1990.

Rothschild, Michael, *Bionomics: Economy as Ecosystem*, Henry Holt, 1990.

Ruttan, Vernon W., *Technology, Growth, and Development: An Induced Innovation Perspective*, Oxford University Press, 2001.

Schelling, Thomas C., "The Cost of Combating Global Warming: Facing the Tradeoffs," *Foreign Affairs*, 76, no. 6 (1997): 8–14.

Schmitz, John E. J., *The Second Law of Life: Energy, Technology, and the Future of Earth As We Know It*, William Andrew, 2007.

Schneider, Anne L. and Helen M. Ingram, *Policy Design for Democracy*, University Press of Kansas, 1997.

Schor, Juliet, "Sustainable Consumption and Worktime Education," *Journal of Industrial Ecology*, 9, no. 1, (2005): 37–50.

Schumacher, E. F., *Small Is Beautiful: Economics as if People Mattered*, Perennial Library, 1989.

Sharpe, Andrew, "Ten Productivity Puzzles Facing Researchers," *International Productivity Monitor*, 9, no. 2, (2004): 15–24.

Simon, Julian L., *The Ultimate Resource 2*, Princeton University Press, 1996.

Sinclair, Matthew, Craig Clauser and Marla Nobles, "The NPT Top 100: An In-Depth Study of America's Largest Nonprofits," *Nonprofit Times*, (November, 2005): 27–41.

Skidelsky, Robert, *Keynes: A Very Short Introduction*, Oxford University Press, 2010.

Skousen, Mark, *The Making of Modern Economics: The Lives and Ideas of the Great Thinkers*, M. E. Sharpe, 2001.

Smith, Adam, *An Inquiry Into the Nature and Causes of the Wealth of Nations*, Clarendon Press, 1976.

Smith, J. B. and D. Tirpak, *The Potential Effects of Climate Change on the United States*, 3 vols., Hemisphere Publishing, 1990.

Sodhi, N. S., L. P. Koh, B. W. Brook and P. K. L. Ng. 2004. "Southeast Asian biodiversity: an impending disaster," *Trends in Ecology and Evolution*, 19, no. 12, (2004): 654–660.

Solow, Robert M., "A Contribution to the Theory of Economic Growth," *Quarterly Journal of Economics*, 70 (1956): 65–94.

———, "Investment and Technical Progress," in K. J. Arrow, S. Karlin and P. Suppes, eds., *Mathematical Models in the Social Sciences*, Stanford University Press, 1960.

———, *Growth Theory: An Exposition*, Oxford University Press, 1970.

———, "The Economics of Resources or the Resources of Economics," *American Economics Review*, 64, no. 2 (1974): 1–14.

Sommers, Albert T., *The U.S. Economy Demystified: What the Major Economic Statistics Mean and Their Significance for Business*, Lexington Books, 1985.

Sorkin, Andrew Ross, *Too Big to Fail: The Inside Story of How Wall Street and Washington Fought to Save the Financial System—And Themselves*, Penguin Books, 2009.

Stannard, David E., *American Holocaust: The Conquest of the New World*, Oxford University Press, 1992.

Stein, Herbert and Murray Foss, *The Illustrated Guide to the American Economy*, 2nd ed., American Enterprise Institute Press, 1995.

Stern, David I., "The Rise and Fall of the Environmental Kuznets Curve," *World Development*, 32, no. 8, (2004): 1419–1439.

Sills, D. L., and R. K. Merton, editors. 2000. *Social Science Quotations: Who Said What, When, and Where*. Transaction Publishers, Piscataway, New Jersey. 437pp.

Théré, Christine and Loïc Charles, "The Writing Workshop of François Quesnay and the Making of Physiocracy," *History of Political Economy*, 40, no. 1 (2008): 1–42.

Tidwell, Mike, *Bayou Farewell: The Rich Life and Tragic Death of Louisiana's Cajun Coast*, Pantheon, 2003.

Tobin, James, "Economic Growth as an Objective of Government Policy," *American Economic Review*, 54, no. 3 (1964): 1–20.

———, *The New Economics One Decade Older*, Princeton University Press, 1974.

Trauger, D. L., B. Czech, J. D. Erickson, P. R. Garrettson, B. J. Kernohan, C. A. Miller, "The Relationship of Economic Growth to Wildlife Conservation," *Wildlife Society Technical Review*, 03, no. 1 (2003): 1–22.

Turner, Graham, *A Comparison of The Limits to Growth with Thirty Years of Reality*, CSIRO Working Paper Series 2008–2009, 2008.

US Environmental Protection Agency, *Promoting Safety for America's Future*, Office of Pesticide Programs, 2002.

Veblen, Thorstein, "The Preconceptions of Economic Science," *Quarterly Journal of Economics*, 14, no. 2 (1900): 240–269.

———, *The Theory of the Leisure Class*, Houghton Mifflin, 1973.

Wackernagel, M., J. Kitzes, D. Moran, S. Goldfinger and M. Thomas, "The Ecological Footprint of Cities and Regions: Comparing Resource

Availability with Resource Demand," *Environment and Urbanization*, 18, no. 1 (2006):103–112.

Ward, Peter and Donald Brownlee, *Rare Earth: Why Complex Life is Uncommon in the Universe*, Springer, 2000.

Warsh, David, *Knowledge and the Wealth of Nations: A Story of Economic Discovery*, W. W. Norton, 2006.

Weintraub, E. Roy, "Neoclassical Economics," *Library of Economics and Liberty*, 1993, econlib.org/library/Enc1/NeoclassicalEconomics.html; accessed 13 August 2010.

Wenzer, Kenneth C. and Thomas R. West, *The Forgotten Legacy of Henry George*, Emancipation Press, 2000.

Whitaker, J.K., *The Early Economic Writings of Alfred Marshall, 1867–1890 Volume 1*, MacMillan, 1975.

Wils, Annababette, "End-Use or Extraction Efficiency in Natural Resource Utilization: Which is Better?," *System Dynamics Review*, 14, no. 2–3 (1998): 163–188.

Winthrop, Michael F., Richard F. Deckro and Jack M. Kloeber, Jr., "Government R&D Expenditures and US Technology Advancement in the Aerospace Industry: A Case Study," *Journal of Engineering and Technology Management*, 19, no. 3–4, (2002): 287–305.

World Commission on Environment and Development, *Our Common Future*, Oxford University Press, 1987.

Yaffee, Steven Lewis, *Prohibitive Policy: Implementing the Federal Endangered Species Act*, MIT Press, 1982.

———, *The Wisdom of the Spotted Owl: Policy Lessons for a New Century*, Island Press, 1994.

INDEX

A
ABC Nightly News, 259–260
absolute advantage, 63–65
Agrarians, 227–228
agriculture
   in classical economics, 53–56, 58–59
   decline, 183–184
   effect of population, 60, 62
   effect on sustainability, 151–152
   industrialization of, 16–18
   irrigation, 6–7
   origins of money and, 184–185, 193
   soil, 8
   trophic levels, 175–177
Alaska National Interest Lands Conservation Act, 244
Allen, Woody, 33
allocation, 143, 144
American apologists, 77, 87
American Museum of Natural History, 213
An Inconvenient Truth, 247
Anasazi, 156
Anglo-Saxon tradition, 77
animal spirits, 189–190
anti-smoking movement, 263–264, 274
Army Corps of Engineers, 279, 280

B
Bakunin, Mikhail, 226
balanced growth, 123, 282–283
bald eagles, 154
Baltimore & Ohio Railroad, 92
banks, 316–317
barter and GDP, 30–31
Besomi, Danielle, 114–115
Beyond Growth (Daly), 138
Bhopal, India, 28
Bhutan, 41
biodiversity, 201
biomass, 178–180
bottled water, 3–4
Bradley, Robert, 146
Bretton Woods, NH, 232
British Petroleum, xiv
brown growth, 196
Bureau of Economic Analysis, 24–25, 104
Bush, George H. W., 26, 246
Bush, George W., 14, 249–250, 261–262, 313–314
Butler, Nicholas Murray, 92

C
California, 6–7
Cambridge Journal of Economics, 114
cap-and-trade policies, 291–300
capital
   definition, 113
   for economic growth, 111–112
   as factor of production, 107–108
   investment in, 117–119
capitalism, 58, 59, 71, 80–81
capital/labor ratio, 118–120, 123–124
Carter, Jimmy, 226
castigation of liquidating class, 268–274
Cato Institute, 138
Center for the Advancement of the Steady State Economy, 256
China, 8, 27–28, 171, 275–276
chlorofluorocarbons (CFCs), 18–19
Clark, John Bates, 76, 90, 91–94, 95, 97

classes of people, 52–53, 55
classical economics
    history of, 51–74
    prices, 77–78
    production function, 86–87
    transition to neoclassical, 76
climate change, 13, 14–15
Clinton, William J., 54, 246–248
Club of Rome, 138
Cobb, John, 39
Colbert, Jean-Baptiste, 55
Cold War, 230–231, 237, 246
Collins, Robert, 227, 229, 235, 239, 240
Columbia University, 92, 97
common sense, 189–190
communism, 70–74
Communist Manifesto (Marx), 71, 72
comparative advantage, 63–66
Competitive Enterprise Institute, 88, 138
conservation, in tribal cultures, 149–157
Conservation Biology, 145
conservatives, 89, 252–253
conspicuous consumption, 259–274
consumption, 24–25, 29, 38–39, 259–274
Continental tradition, 77
co-production, 37–38
Corn Laws, 55
corporations, 30, 245, 248–249
The Corruption of Economics (Gaffney, Harrison), 85–100
Council of Economic Advisors (CEA), 233–234, 236
critical theory, 71

D
Daly, Herman, 33, 39, 138, 140–142, 157, 172, 183, 217, 299
Das Kapital (Marx), 72, 73
La Décroissance, xv, 291–292
Deepwater Horizon oil spill, xiv
Defense Production Act, 234
Deffeyes, Kenneth, 10–11

Denison, Edward F., 217–218
Dietz, Rob, 305
diminishing returns, 36–37
distribution of wealth, 143, 148–155, 306–310
Domar, Evsey, 109
Ducks Unlimited, 213

E
Earth, 159–161
Earth in the Balance (Gore), 246
Easter Island, 155–156
ecological economics
    origins of, 137–142
    overview, 142–169
Ecological Economics (Daly, Farley), 299
ecological footprints, 186–193, 264–265, 268–270, 308
ecological macroeconomics, 311
The Economic Consequences of the Peace (Keynes), 105
economic growth
    alternatives to, 277–278
    causes of, 31
    consumption and, 29
    corruption of, 87–88
    effect of, 21–22, 32–43
    environmental protection and, 197–205, 208–209, 219–222
    as goal, xiii–xv, 278–281
    limits to, 137
    political history of, 225–257
The Economic Hit Man (Perkins), 232
Economic Journal, 109
economic man, 44, 60, 149
economics
    history of classical, 51–74
    history of neoclassical, 75–116
    overview of ecological, 137–169
    principles of, 36–38
    study of, 41–47
    theories of, 115–116
economies of scale, 217–218
economy of nature
    capacity of, 180–182

GDP of, 177–180
  trophic levels, 173–175
ecosystems
  human economy size and, 143
  trophic levels, 173–175
Ederer, Rupert, 184
education, 125–126
Einstein, Albert, 162
Eisenhower, Dwight D., 41, 226, 236–237, 248
electrical blackouts, 11–13
Elements of Pure Economics (Walras), 79–80
Ellison, Katherine, 172–173
Ely, Richard T., 90, 94–97
emergy, 165–169
Employment Act, 281–285, 319
endangered species, 164, 324
Endangered Species Act, 197–198, 240, 249–250, 280–281, 285–290
endogenous growth theory, 127–134
end-use innovation, 206, 207–208
energy, 159, 161–168
Enough is Enough (Dietz, O'Neill), 290
entropy law, 180
The Entropy Law and the Economic Process (Georgescu-Roegen), 138, 139–140
environmental Kuznets curve, 200–205, 240
environmental protection
  conflict with economic growth, xiv–xv, 197–205, 208–209, 219–222
  R&D, 213–216
  See also Endangered Species Act
The Evolution of Money (Ederer), 184
explorative innovation, 206
extractive innovation, 206

F
factors of production, 58, 83, 84–85, 99–101, 107–108, 111–112, 158
Farley, Joshua, 299
Feder, Kris, 85

Federal Reserve Bank of Kansas City, 315
fee-service banking, 317
Ferguson, Niall, 171
fertilizer, 8
finances. *See* money
food, 6–7, 14, 16–18
For the Common Good (Daly, Cobb), 138, 142
Ford, Henry, 316
Foreign Affairs, 183
fossil fuel cap policy, 291–300
fractional reserve requirements, 316–317
Friends of the Earth, 240
Full Employment and Balanced Growth Act, 281–285, 319
Full Seas Act, 319

G
G100 nations, 294–296
Gaffney, Mason, 85–100
Galbraith, John Kenneth, 63, 238–239
garbage, 28–29
GDP (gross domestic product)
  ecological footprint and, 188, 189–193, 269
  effect of population, 29–30, 132–134
  factors affecting, 123
  as measure, 25–26
  in nature, 177–180
  quality of life and, 38–40
  R&D in, 215
  supplementation of, 322–326
  technological progress and, 122
general equilibrium, 78–79
The General Theory of Employment, Interest, and Money (Keynes), 106–108, 229, 234, 282
genetically modified crops, 17
Genuine Progress Indicator (GPI), 323–324
George, Henry, 56, 74, 75, 80–96
Georgescu-Roegen, Nicolas, 138, 139–140

Germany, 105, 233
Gilman, Daniel Coit, 94–96
Glennon, Robert, 5
global economy, xiv–xv, 63–66
Global Footprint Network (GFN), 36
global warming, 13, 14–15
GNP (gross national product), 26, 231, 234
gold, 168
Goodland, Robert, 142
goods, 23, 130
Gore, Al, 246–247
Gossen, Hermann Heinrich, 90
government
    democratic, 73–74
    expenditures, 30
    role in financial system, 105–106
    See also politics
Great Depression, 105, 228–229, 282, 321–322
Great Society, 238–239
green growth, 196–198
greenhouse gases, 13
Gross, Daniel, 266
Gund Institute, 311

H
Hall, Charlie, 165
Happy Planet Index (HPI), 323–324
Harrison, Fred, 85
Harrod, Roy, 109–115
Harrod-Domar model, 109, 117
Hebrew tribes, 152
Heilbroner, Robert, 58, 62, 66
Holmgren, David, 159
Homo ecologicus, 154–155
Homo economicus, 44, 60, 149
Howard Baker, 287–288
Hubbert, Marion King, 10
Hubbert's Peak (Deffeyes), 10
human behavior, 149–155
human capital, 125–126

human economy
    capacity of, 182–183
    trophic levels, 175–178
Hummer haters, 243–244, 259–260

I
immigration, 83, 301
Index of Sustainable Economic Welfare (ISEW), 39–40
India, 8, 12–13, 28
indicators, 322–326
Industrial Revolution, 153–154
inflation, 188, 242
information economy, 171–173
Ingram, Helen, 281
innovation, 205–208
input-output analysis, 204–205
An Inquiry into the Nature and Causes of the Wealth of Nations (Smith), 57, 184
Intergovernmental Panel on Climate Change, 13
International Monetary Fund (IMF), 108, 231–232, 235
International Workingmen's Association, 226
Introduction to Economic Growth (Jones), 125
Iron Law of Wages, 63, 68
iron triangles, 248–249
irrigation, 6–7

J
Japan, 26, 233
Jevons, William Stanley, 78
Johns Hopkins University, 92, 95, 97
Johnson, Alvin S., 90
Johnson, Lyndon Baines, 238–239, 241
Jones, Aaron, 202
Jones, Charles I., 125, 132–133

K
Kantrow, Yvette, 268
Kennedy, John F., 237–239

Index | 363

Keynes, John Maynard, 104–109, 115, 184, 189, 229, 282
Keyserling, Leon H., 233–234
Khrushchev, Nikita, 237
Knight, Frank, 90, 97–100
Krugman, Paul, 208
Kyoto Protocol, 300

L
La Décroissance, xv, 291–292
labor
   conditions, 66
   oppression of, 80–81
   role in economic production, 58, 117–124
   wages, 62–63, 68
   See also Employment Act
labor intensity, 320–321
LaDuke, Winona, 247
laissez faire, 56–57
land
   as capital, 97, 99, 113–114
   in production function, 87–88, 93, 112, 116, 158
land tax, 81, 82–84
landfills, 28–29
land-grant schools, 94–95
Leonhardt, David, 250–252
Leontief, Wassily, 79, 204–205
Leopold, Aldo, 69, 270
Limits to Growth (Meadows, Meadows, Randers), 138–139
liquidating class, 264–274, 294–296
local currencies, 317–318
Lomborg, Bjorn, 88, 145
Louisiana, 15
Lucas, Robert E., 125, 127
Luzzati, Tommaso, 133–134

M
macroeconomics, 46–47, 142–143, 227
Malaysia, 208
Malthus, Thomas Robert, 59–62, 74, 142

Mankiw, Gregory, 125
manufacturing sector, 18–20, 177, 199
marginal utility, 77–78
Marginalist Revolution, 77–80
Marshall, Alfred, 76, 100–103, 168
Marx, Karl, 71–74, 75, 80–81, 167, 226
Maslow, Abraham, 270–271
McCloskey, Deirdre, 125
McNeil, J. M., xiii
Meadows, Dennis, 138
Meadows, Donella, 138, 139–140, 326
Menger, Carl, 76, 78
Mesopotamia, 186–187
methane, 13
Michigan State University, 204, 208
microeconomics, 44–46, 100–101, 142–143
Mill, John Stuart, 66–71, 90, 142
Missoula, Montana, 3–4
MIT Dictionary of Modern Economics, 113
money
   flow of, 157–159
   policies, 311–318
   saving, 122
   trophic theory of, 184–193, 195, 269
moon, 159–161
More (Collins), 227
Morrill Act, 94–95

N
Nader, Ralph, 246–247
Naisbitt, John, 171
Natanes Plateau, Arizona, 4–5
national economies, xiv–xv, 26, 86
National Football League, 307–308
national income accounting, 24–25, 104
National Science Foundation, 210, 213
natural capital
   as base of economy, 176–178
   biomass, 178–180
   cap-and-trade, 296–298

human economy draw of, 198–199
  scarcity of, 144–148
The Nature Conservancy, 213
neoclassical economics
  environment and, 143
  history of, 54, 75–116
  the market, 144
  money flow, 157–159
neoconservatives, 32
neoliberals, 32
New Age philosophy, 159–160
New Deal, 106, 229–230, 321
New Economic Policy, 242
New Frontier, 238–239
New York Times, 13, 266
New York Times Magazine, 250
9/11 terrorist attack, 261–262
nitrous oxide, 13
Nixon, Richard M., 241–243
non-governmental organizations
  (NGOs), 237, 240
non-profit R&D, 213–214
Nordhaus, William, 183
Norgaard, Richard, 145–146
NSC-68, 231, 234
nuclear technology, 163–164

O

Obama, Barack, 14, 250–251, 253–257
Odum, Howard T., 165–169
oil, 8–11, 291–300
Olduvai Theory of Energy Production, 11
Organization of Petroleum Exporting
  Countries (OPEC), 10
Ormerod, Paul, 73
Outlines of Economics (Ely), 94
ozone, 18–19
Ozone Crisis (Roan), 19

P

patents, 130–131
Paulson, Henry, 313–315
Pearce, David W., 114

per capita consumption, 29, 36
Perkins, John, 232
physiocrats, 52–59, 82, 86–87
Poland, 152
policies
  cap-and-trade, 291–300
  development in academia, 310–311
  discussion of options, 276–277
  distribution of wealth, 306–310
  economic growth as goal, 278–281
  employment, 281–285, 318–322
  endangered species, 285–290
  measurement of progress, 322–326
  monetary, 311–318
  population stabilization, 300–306
politics
  economics and, 54–55
  history of economic growth,
    225–257
population
  as consumers, 29
  for economic growth, 112
  effect of growth, 32–35, 52, 60
  employment and, 282
  GDP growth and, 132–134
  green growth, 197
  stabilization, 68–70, 300–306
Price, Richard, 59
prices, 77–78, 167–168
Principles of Economics (Marshall),
  91, 102
The Principles of Political Economy
  (Mill), 67–71
process innovation, 206
product innovation, 206
production, 24–25
production function, 86–88, 93, 99,
  158
production model, 117–118
Progress and Poverty (George), 80,
  82–83
prosperous way down, 166–167
A Prosperous Way Down (Odum),
  165

Q

Quesnay, Francois, 51–57, 175–176, 204

R

Randers, Jorgen, 138
Reagan, Ronald, 244–246, 279–280
Redefining Progress, 40
Reffalt, Bill, 286
refrigeration industry, 18–19
religions, 152–153
rent, 81–82
research and development (R&D), 129–132, 209–216, 218, 219–222
Ricardo, David, 60–66, 68, 82, 167
rivalry, 130
Roan, Sharon, 19
Rockefeller, John D., 92
Rockefeller Brothers Fund, 237
Rolston, Holmes III, 280–281
Romer, David, 125
Romer, Paul, 127, 134, 209
Roosevelt, Franklin D., 106, 228
Rostow, Walt Whitman, 70, 102, 218
Rothschild, Mayer Amschel, 316
Russian Revolution, 226–227
Ruttan, Vernon, 212, 215–216

S

salary caps, 307–308
Samuelson, Paul, 76
San Carlos Apache Tribe, 4–5, 202–203
S-A-T-G model, 278, 280, 281
Say, Jean Baptiste, 24
Say's Law, 24
scale, 143–144, 155
Schelling, Thomas C., 183–184
Schlesinger, Arthur M., 239
Schneider, Anne, 281
Schumacher, E. F., 217
Schwarzman, Stephen A., 265–270
self-sufficient services fallacy, 172–173
Seligman, Edwin R. A., 90, 92–94
services
    definition, 23
    natural capital use, 196, 199–200
    rivalrous, 130
    trophic levels, 177
Shoveling Fuel for a Runaway Train (Czech), 114, 124, 172, 260–261, 264, 271
Simon, Julian, 145
Sinclair, Upton, 90
single-tax movement, 82–84, 89, 93, 99
The Skeptical Environmentalist (Lomborg), 88, 145
slavery, 273–274
Smith, Adam, 57–59, 144, 167, 184
snail darter, 287–288
soil, 7–8
Solow, Robert, 117–125
Sorkin, Andrew Ross, 312
Southern Pacific Railroad, 92
Soviet Union, 230–231, 233
spring water, 4
stagflation, 241, 244
Stanford University, 92
State of the World (Worldwatch Institute), 20
stationary state, 67–70
steady state, 119–120, 123–124
steady state economy
    definition, 119–120
    demand for, xiv–xv
    distribution of wealth, 306–310
    employment policies, 281–285, 318–322
    endangered species policies, 285–290
    as goal, 278
    measurement of, 322–326
    monetary policies, 311–318
    movements, 291–292, 294–296
    policy framework, 290–291
    policy options, 276–277
    population stabilization, 300–306
    in religion, 152–153
steady state revolution, 260–274

Summers, Lawrence, 255
sun, 159–161
supply and demand, 144–148

T
Tableau Economique (Quesnay), 52–54, 204
taxes, 55–57, 81, 83–84, 308
technological progress
  ecological footprint, 189
  for economic growth, 111–112, 121–122
  green growth, 197
  innovation, 205–208
  See also research and development
Tellico Dam, 287–288
Tennessee Valley Authority, 287–288
Thailand, 41
Thanksgiving, 154
thermodynamics, laws of, 161–163, 180
Third World, 235–236
Tobin, James, 238
Too Big To Fail (Sorkin), 312
transition path, 123
Treatise on Political Economy (Say), 24
tribal cultures, 149–157
trophic levels
  economy of nature, 173–175
  human economy, 175–178
  perpetual growth and, 180–182
trophic theory of money, 184–193, 195, 269
Truman, Harry, 234
Trust for Public Land, 214

U
unemployment, 242
United Arab Emirates, 294
United States
  history of economic growth politics, 225–257
  measurement of economy, 26–27
  national income accounting, 24–25, 104

pesticide use, 16–17
political system, 73–74
R&D, 209–210
soil erosion, 8
US Fish and Wildlife Service, 287
US Forest Service, 279
US National Climatic Data Center, 13
US National Security Council, 231
universities, 94–95, 97–98, 310–311
University of California Press, 261
University of Chicago, 92, 97–98
unlimited wants, 45
Unquenchable (Glennon), 5

V
value, 167–168
Valuing the Earth (Daly), 138
Veblen, Thorstein, 76
Vietnam War, 240–241

W
wages, 62–63, 68
Walker, Francis A., 93–94, 95
Wall Street Journal, 256
Wallace, William J., 90
Walras, Léon, 78, 79–80, 90
War on Poverty, 239
water, 3–7, 14
wealth, distribution of, 143, 148–155, 306–310
The Wealth of Nations (Smith), 57, 184
Weigley, Russell, 231, 233
Wen Jiabao, 276
Wilderness Act, 244
Wildlife Conservation Society, 213
Will, George, 145
World Bank, 108, 142, 231–232, 235
World Trade Center attack, 261–262
World War I, 104–105
World War II, 230–232
Worldwatch Institute, 20

Y
Yaffee, Steven, 286

ABOUT THE AUTHOR

Brian Czech is the founding president and executive director of the Center for the Advancement of the Steady State Economy (CASSE), based in Arlington, Virginia. Czech wrote *Supply Shock* from 2000-2013, a period encompassed by his career at U.S. Fish and Wildlife Service headquarters (1999-2017). There, he was prohibited from writing and presenting on the conflict between economic growth and wildlife conservation, a central conflict of his Ph.D. research at the University of Arizona in the 1990s. Czech resigned from the government in 2017 to run CASSE full time. His other titles include *Shoveling Fuel for a Runaway Train: Errant Economists, Shameful Spenders, and a Plan to Stop Them All* (2000, University of California Press), and *The Endangered Species Act: History, Conservation Biology, and Public Policy* (2001, Johns Hopkins University Press). *Supply Shock* was originally published by New Society Publishers (2013).

If you have enjoyed *Supply Shock* you might also enjoy other
STEADY STATE PRESS TITLES

Best of The Daly News
Uncommon Sense

Steady State Press is an imprint of CASSE
CASSE MISSION STATEMENT

The mission of CASSE is to advance the steady state economy, with stabilized population and consumption, as a policy goal with widespread public support. We pursue this mission by:

- educating citizens, organizations, and policy makers on the conflict between economic growth and (1) environmental protection, (2) ecological and economic sustainability, and (3) national security and international stability;
- promoting the steady state economy as a desirable alternative to economic growth;
- studying the means to establish a steady state economy.

Steady State Press
steadystate.org